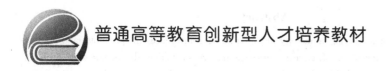
普通高等教育创新型人才培养教材

Powder Materials and Technologies in Aeronautics and Astronautics

航空航天粉末材料与技术

Zhang Shasha　　　Oleksandr Moliar　　　Yao Zhengjun

张莎莎　　〔乌克兰〕亚历山大·莫利亚尔　　姚正军　　主编

北京航空航天大学出版社

Abstract

As a basic textbook for industrial technology majors, this book introduces the principles and methods of material selection and processing of typical aerospace materials, taking the engineering problems related to typical materials in the fields of aviation, aerospace and civil aviation as the starting point. It also systematically describes the basic principles and technologies of powder metallurgy and introduces new theories and technologies such as additive manufacturing and spray deposition. The application of powder metallurgy technology to typical components in aerospace is clarified.

This textbook can be used as a textbook for undergraduate students majoring in materials science and engineering, mechanical engineering or other related majors. It can also be used as a tool for graduate students and teachers, as well as a training book and reference book for engineers and technicians engaged in the aerospace engineering and other fields.

图书在版编目(CIP)数据

航空航天粉末材料与技术 ＝ Powder Materials and Technologies in Aeronautics and Astronautics : 英文 / 张莎莎,(乌克兰)亚历山大·莫利亚尔(Oleksandr Moliar), 姚正军主编. -- 北京 : 北京航空航天大学出版社,2022.10

ISBN 978-7-5124-3901-6

Ⅰ. ①航… Ⅱ. ①张… ②亚… ③姚… Ⅲ. ①航空材料－粉末冶金－金属材料－研究－英文②航天材料－粉末冶金－金属材料－研究－英文 Ⅳ. ①V25

中国版本图书馆 CIP 数据核字(2022)第 176543 号

版权所有,侵权必究。

Powder Materials and Technologies in Aeronautics and Astronautics
航空航天粉末材料与技术
Zhang Shasha　　　Oleksandr Moliar　　　Yao Zhengjun
张莎莎　　〔乌克兰〕亚历山大·莫利亚尔　　姚正军　　主编
策划编辑　董 瑞　　责任编辑　宋淑娟

*

北京航空航天大学出版社出版发行

北京市海淀区学院路 37 号(邮编 100191)　http://www.buaapress.com.cn
发行部电话:(010)82317024　传真:(010)82328026
读者信箱:goodtextbook@126.com　邮购电话:(010)82316936
北京建宏印刷有限公司印装　各地书店经销

*

开本:710×1 000　1/16　印张:15.75　字数:345 千字
2022 年 10 月第 1 版　2022 年 10 月第 1 次印刷　印数:1 000 册
ISBN 978-7-5124-3901-6　定价:59.00 元

若本书有倒页、脱页、缺页等印装质量问题,请与本社发行部联系调换。联系电话:(010)82317024

Preface

  Creation of new industries and development of traditional ones are determinatively associated with the materials. For a good reason the name of epochs of civilization development derives from the name of materials— stone, bronze, and iron ages; only later did sociology interfere. For a new tool, weapon, device, machine, etc. to appear, a person should have an appropriate material base for their production, and then the constructor's talent, craftsman's skills, technical and technological skills, organization and tooling backup come into action.

  Considering fast developments, the aviation shows the most visible mutual influence of materials and level of development of technology. If to analyze the design of aircrafts of the beginning of the aviation age and the modern aircrafts, their technical capabilities are fundamentally different. These differences are primarily in the intensity of perceived external loads. This became possible only due to the development of science of materials, new materials, the appearance of which was, of course, determined by the needs of technology development, in our case, of aviation. The airframe of the first aircraft was known to be made mostly from wood (plywood) and fabric. The reciprocating internal combustion engines were made from non-alloy steel and cast iron. Over time, aluminum and magnesium, and then titanium alloys, were used in structures, and heat-resistant alloy steels and nickel alloys were used in engines. When came to the age of composite materials on metal and polymer matrix, new approaches were developed in structural design of new materials. Thus, we observe that the rapid growth of aviation capabilities over a rather short historical period (100—150 years) has caused qualitative changes in life on Earth, providing unprecedented possibilities for people's communication, human's entry into outer space, introduction of developed technologies to other industries. One of the areas of science of materials of aircraft manufacturing is powder metallurgy, which provides new, previously unattainable opportunities for obtaining materials with unique properties, which will be discussed in this book.

  This book mainly focuses on the application of powder metallurgy technology in aerospace field. The feasibility and superiority of powder metallurgy are presented through the study of existing powder metallurgy technologies and materials in aerospace field, and the general preparation methods of powder metallurgy products are introduced. On the basis of introducing the basic principle of powder metallurgy, the main material categories such as powder metallurgical structural materials, composite materials, friction materials, and functional materials are combined with the applications of key aerospace components to provide a comprehensive description of powder materials and technologies for aerospace use. The construction of aerospace powder materials and technology has a far-reaching impact on the cultivation of talents urgently needed in the field of aerospace, and will play an important role in promoting the development of "materials science and engineering specialty" with the characteristics of aerospace and national defense and the construction of world-class disciplines.

The second, fourth, sixth, ninth chapters and designations and abbreviations of this book were written by Shasha Zhang, the first, third, fifth and seventh chapters were written by Oleksandr Moliar and modified by Shasha Zhang, and the eighth chapter was written by Zhengjun Yao. Most examples and data of this book were provided by Oleksandr Moliar. The whole book was drafted and finalized by Shasha Zhang and reviewed by Zhengjun Yao.

In the process of writing this book, it has received strong support from the School of Materials Science and Technology of Nanjing University of Aeronautics and Astronautics, Key Laboratory of Materials Preparation and Protection for Harsh Environment (Nanjing University of Aeronautics and Astronautics), Ministry of Industry and Information Technology and the Academic Affairs Office of Nanjing University of Aeronautics and Astronautics. We are deeply grateful for their supports. At the stage of reviewing, colleagues and experts from many sister colleges and publishing houses have put forward many valuable opinions on the compilation of the outline and text, and we would like to express our sincere thanks.

Due to the limited level of the editors, there are inevitably inappropriate things in the book, and we urge readers to criticize and correct.

Editors
Jan. 8th, 2019

Designations and Abbreviations

Designation	Property (Term)	Unit of measurement
m	Mass	kg
ρ	Density	kg/m^3
ρ_{th}	Theoretical density	kg/m^3
T	*Temperature*	℃
T_{mel}	Melting temperature	
T_{ev}	Evaporation temperature (boiling point)	
T_{rec}	Recrystallization temperature	
T_{pt}	Temperature of polymorphic transformation	
T_{sin}	Sintering temperature	
T_h	Harding temperature	
T_{en}	Aging temperature	
T_{br}	Brazing temperature	
T_{dis}	Dissociation temperature	
HV	Vickers hardness	MPa
HB	Brinell hardness	
H_μ	Microhardness	
HRA	Rockwell hardness A	—
HRC	Rockwell hardness C	
HRF	Rockwell hardness F	
R_m	Tensile strength	MPa
$R_{p0.2}$	Yield strength	
$R_{m,ben}$	Bending strength	
$R_{m,com}$	Compressive strength	
R_m^T	Ultimate tensile strength at temperature T	
R_t^T	Long-term strength at T and given time t	
$R_{\varepsilon/t}^T$	Creep limit at T: ε — deformation (%), t—time (h)	
τ_{sh}	Resistance strength to shear	
R_m/ρ	Specific strength	m

(continued)

Designation	Property (Term)	Unit of measurement
E	Modulus of normal elasticity (Young's modulus)	MPa
E_d	Dynamic modulus of normal elasticity	
E_{com}	Modulus of elasticity at compression	
G	Elastic modulus at shear (shear modulus)	
E/ρ	Specific modulus of elasticity	m
A	Elongation	%
Z	Section shrinkage	%
KCU	Impact strength, sample with U-shape notch	J/m²
KCV	Impact strength, sample with V-shape notch	
KCT	Impact strength, sample with a crack	
K_{1c}	Critical stress intensity factor	MPa $\sqrt{m}$
K_{1scc}	Critical stress intensity factor with corrosion cracking	
α	Thermal coefficient of linear expansion (LCTE)	1/K
R	Specific electrical resistivity	$\Omega \cdot m$
ΔH_{mel}	Specific heat of fusion	kJ/kg
ΔH_{ev}	Specific heat of vaporization	
Q_1	Specific heat of combustion of metal	
Q_2	Metal oxide formation enthalpy	
μ	Friction coefficient	—
t	Time	s
v	Velocity	m/s
v_{kr}	Crystallization velocity	—
P	Pressure	MPa
P_{pr}	Pressing pressure	
d	Diameter	mm
h	Height	mm
PV	Qualitative criterion for evaluating the work of the friction pairs	MPa $\cdot$ m/s
p	Porosity	%
c	Specific heat	kJ/(kg $\cdot$ K)
λ	Coefficient of thermal conductivity	W/(m $\cdot$ K)
B_r	Residual magnetic flux density	T
H_c	Coercive force	A/m

(**continued**)

Designation	Property (Term)	Unit of measurement
$(BH)_{max}$	Maximum energy product	kJ/m^3
ε	Dielectric constant	—
$\tan \delta$	Dielectric loss tangent	
ρ_b	Apparent density	kg/m^3
REM	Rare Earth Metals	—
SPS	Spark Plasma Sintering	
GTE	Gas Turbine Engine	
CE	Coefficient of Efficiency	

Contents

Chapter 1　Basic Materials for Aerospace Engineering …… 1
　1.1　Basic Metallic Materials Used in Aerospace Engineering …… 1
　1.2　Approaches to Choosing of Structural Materials for Aircraft …… 2
　1.3　Some Examples of Choosing Materials for Specific Airplane Components …… 5
　1.4　Role of Materials in Aerospace Engineering …… 8

Chapter 2　Introduction to Powder Metallurgy …… 17
　2.1　The Properties of Powders …… 19
　　2.1.1　Physical Properties …… 19
　　2.1.2　Chemical Properties …… 23
　　2.1.3　Processing Properties …… 24
　2.2　Producing Sintered Products …… 25
　　2.2.1　Powder Preparation …… 25
　　2.2.2　Powder Forming …… 26
　　2.2.3　Sintering …… 31
　2.3　Additional Processes After Sintering …… 33

Chapter 3　Basic Principles of Powder Metallurgy of Nanomaterials …… 36
　3.1　General Information …… 36
　3.2　Methods of Obtaining Nanopowders …… 39
　3.3　Size Effects in Nanoparticles …… 41
　3.4　Consolidation of Bulk Nanomaterials …… 43
　　3.4.1　Pressing of Nanopowders …… 44
　　3.4.2　Sintering …… 48

Chapter 4　Structural Materials …… 52
　4.1　Powder Metallurgy Aluminum Alloys …… 52
　4.2　Powder Metallurgy Titanium Alloys …… 58
　4.3　Powder Metallurgy Magnesium Alloys …… 63
　4.4　Powder Metallurgy Beryllium Alloys …… 65

Chapter 5 Heat Resistant Materials and Materials Strong at High Temperatures ... 68
- 5.1 Refractory Metals and Alloys ... 69
- 5.2 Superalloys ... 73
- 5.3 Dispersion-strengthened Material ... 80
- 5.4 Refractory Metal Compounds (Intermetallic Compounds) ... 88
- 5.5 Special Ceramic Materials with High Temperature Strength ... 90

Chapter 6 Composite Materials ... 94
- 6.1 General Concepts About Composite Materials ... 94
- 6.2 Methods of Obtaining Composite Materials ... 96
- 6.3 Metallic Structural Composite Materials ... 97
- 6.4 Heat-resistant Composite Materials ... 101

Chapter 7 Materials for Friction Joints ... 109
- 7.1 Introduction ... 109
  - 7.1.1 Basic Terms of Tribology ... 110
  - 7.1.2 Some Examples of the Influence of Friction on the Operation of Mechanics ... 114
- 7.2 Bearing Powder Materials ... 116
  - 7.2.1 Iron-based Bearing Materials ... 119
  - 7.2.2 Copper-based Bearing Materials ... 120
  - 7.2.3 Metal-polymer Bearing Materials ... 123
  - 7.2.4 Nickel-based Bearing Materials ... 127
- 7.3 Friction Powder Materials ... 129
  - 7.3.1 Copper-based Friction Materials ... 131
  - 7.3.2 Iron-based Friction Materials ... 132
  - 7.3.3 Cermets for Aircraft Brakes ... 134
  - 7.3.4 Composite Materials with Carbon and Ceramic Matrix ... 135

Chapter 8 Special-purpose Functional Materials ... 140
- 8.1 Concept of Functional Materials, Their Role in Engineering ... 140
- 8.2 Heavy Alloys ... 141
- 8.3 Porous Permeable Materials ... 143
  - 8.3.1 Filters ... 144
  - 8.3.2 Porous Ionizer ... 148

8.3.3	Transpiration Cooling Materials	149
8.3.4	Capillary Structures of Heat Pipes	154
8.3.5	Porous Electrodes for Fuel Cells	156

8.4 Sealant Materials ·· 156

8.5 Powder Materials for Electrical and Radio Technical Purposes ········· 159

8.5.1	Optical Ceramics	160
8.5.2	Radio-wave Transparent Materials	167
8.5.3	Magnetic Materials	174

Chapter 9 New Technologies in Powder Metallurgy ···················· 179

9.1 Nanomaterials and Nanotechnologies ·· 179

9.2 Additive Manufacturing ·· 194

9.3 Severe Plastic Deformation of Sintered Billets ··························· 208

9.4 Self-propagation High-temperature Synthesis ······························ 211

9.5 Metal Injection Molding ··· 221

References ·· 227

Chapter 1　Basic Materials for Aerospace Engineering

1.1　Basic Metallic Materials Used in Aerospace Engineering

One modern aircraft (airplane, helicopter, rocket) uses hundreds of brands of materials of various functional purposes. Leading global aircraft manufacturers in different periods of time show different ratios between the materials used for producing subsonic airplanes, but these ratios are not fundamentally different. Fig. 1.1 presents the ratio between the main structural materials in the aircrafts of Antonov company designed during different periods.

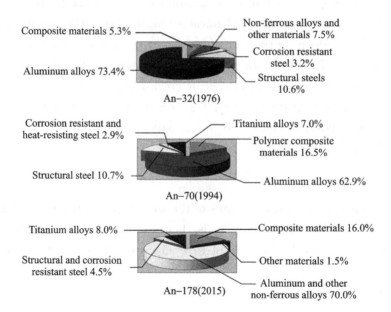

FIGURE 1.1　Basic materials used in the aircrafts in different periods of their production (in brackets— the year of the first flight)

The main structural material in aircraft construction (65%—75%) has been aluminum and its alloys for many years. In addition to aluminum, structural (medium and high strength) and corrosion resistant steels (8%—10%) and titanium alloys (3%—10%) take a prominent position in aircraft. In recent years, the part of magnesium alloys in aircraft structures, due to their low corrosion resistance and increment in the aircraft's calendar life, has decreased, with the amount of titanium alloys increased for the same reason. The volume of composite materials on a non-metallic and, to a lesser extent, a metallic matrix has increased quite substantially. According to all aircraft companies, the weight of composite materials is up to 20% of the weight of the airframe, replacing the aluminum alloys. According to the information of Boeing company, the carbon-epoxy composite weight makes 50% of the weight of the airframe of Dreamliner B787, which is currently a record in the use of composite materials[1]. The increase in the use of composite materials led in turn to increase in the application of titanium alloys, in particular, for fasteners, because of its insensitivity to contact corrosion.

1.2 Approaches to Choosing of Structural Materials for Aircraft

When choosing a material for a specific part of an aircraft, first, one should pay attention to ensuring its mechanical strength and stiffness with a minimum mass, that is, to ensuring the maximum weight efficiency of the material.

The weight efficiency of the material is estimated based on the specific strength R_m/ρ, specific stiffness E/ρ, and specific crack resistance K_{1c}/ρ.

From a physical point of view, The concept of specific strength determines length of this material which fails by gravity of Earth without loading and is measured in meters:

$$\frac{R_m}{\rho} = \frac{\text{kg/mm}^2}{\text{kg/m}^3} = 10^6 \text{ m} \qquad (1.1)$$

Analyzing the working conditions of the aircraft parts, it can be concluded that most of them work under bending loads. If to compare two parts made of different materials at such a load, the ratio of their masses will be based on the following conditions:

- strength condition:

$$\frac{m_1}{m_2} = \left(\frac{R_{m_2}}{R_{m_1}}\right)^{\frac{2}{3}} \times \frac{D_1}{D_2} \qquad (1.2)$$

- stiffness condition:

$$\frac{m_1}{m_2} = \left(\frac{E_1}{E_2}\right)^{\frac{1}{2}} \times \frac{D_1}{D_2} \qquad (1.3)$$

and the ratio of cross sections of components that perceive this bend will be:

$$\frac{\delta_2}{\delta_1} = \left(\frac{E_1}{E_2}\right)^{\frac{1}{4}} \qquad (1.4)$$

where m_1 and m_2—masses;

D_1 and D_2—diameters;

E_1 and E_2—elastic modulus;

δ_1 and δ_2—cross section of the first and second components, respectively.

The generalized comparative properties of structural materials for aerospace (at temperature of 20 ℃) are given in Table 1.1[2].

TABLE 1.1 Comparative properties of materials used in aircraft constructions

Materials	Density $\rho/(\text{kg} \cdot \text{m}^{-3})$	Tensile strength R_m/MPa	Elastic modulus E/MPa	Specific strength $\dfrac{R_m}{\rho}/\text{km}$	Specific stiffness $\dfrac{E}{\rho}/\text{km}$
Aluminum alloys	2 700	400—650	72	14.8—24.0	26 500
Magnesium alloys	1 800	200—340	45	11.0—18.9	25 000
Titanium alloys	4 500	500—1300	120	11.0—29.0	26 600
Medium strength steels	7 800	800—1 300	210	10.3—16.7	27 000
High strength steels	7 800	1 300—2 300	210	16.7—29.5	27 000
Composite materials	1 400—2 600	500—1 300	35—250	40—60	25 000—100 000

Considering that modern aircraft (including supersonic), as well as engines and rockets operate at elevated temperatures, it is often necessary to take into account the properties of materials at high temperatures. The dependence of the strength property on the service temperature of the main structural materials is shown in Fig. 1.2.

A more visual representation of the effect of temperature on the properties of materials is demonstrated by the dependence of "specific strength – temperature" (see Fig. 1.3). It is indicated that even a low-alloyed ($\alpha+\beta$) -titanium alloy (TC4) exceeds both aluminum alloys and steels to a temperature of 150 ℃ in terms of the specific strength. In the temperature range of $T = 300—500$ ℃, the TC18 high-alloyed high-strength titanium alloy possesses the maximum specific strength whereas the specific strength of TC4 titanium alloy gradually decreases. The evolution of specific strength as a function of temperature implies that titanium alloys are more efficient than other structural materials for the application in aircraft.

The specific strength – temperature dependence also indicates that starting from a certain temperature value (different with different types of materials), a

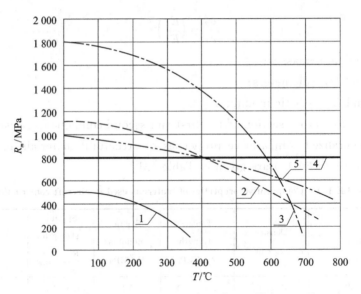

1—Aluminum alloys; 2—Medium-strength steels; 3—High-strength steels;
4—Nickel superalloys; 5—Titanium alloys

FIGURE 1.2 Evolution of strength properties (R_m) of basic structural materials depending on the temperature (T)

sharp decrease is observed in the specific strength, which must be considered when designing the construction. In addition, for materials used at elevated temperatures, it is also necessary to consider the high-temperature creep. The creep behavior of the materials can lead to the change of the size of the part under constant load, which is much lower than the yield tensile strength of the material at the test temperature.

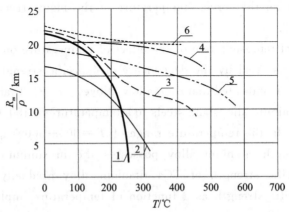

1—Aluminum alloy 7075; 2—Aluminum alloy 2024; 3—Titanium alloy TC4;
4—Alloyed structural steel 30CrMnSi;
5—Corrosion resistant steel 08Cr15Ni5Cu2Ti; 6—Titanium alloy TC18

FIGURE 1.3 Dependence of specific strength (R_m/ρ) of structural alloys on temperature (T)

1.3 Some Examples of Choosing Materials for Specific Airplane Components

In order to understand approaches to choosing materials for the airplane's airframe, it is necessary to know how its basic parts work. Fig. 1.4 demonstrates the airplane structure of Boeing 787 Dreamliner[3]. The plane's airframe can be classified into the following main parts:

① fuselage (which, in turn, is divided into nose, center and tail parts);
② wing;
③ vertical and horizontal stabilizers (aircraft's empennage);
④ landing gear;
⑤ power plant;
⑥ systems (aircraft control, hydraulic, fuel, air).

Each of these parts perceives its own specific (always alternating) loads, which primarily determine the choice of material for their manufacture.

FIGURE 1.4 Airplane structure of Boeing 787 Dreamliner

The fuselage is traditionally made of aluminum alloys: the cover and longitudinal frameworks are made from alloy 2024 and its modified alloys (Al-Cu-Mg alloying system), and transverse framework is from high-strength alloys

7075, 7055 (Al-Zn-Mg alloying system). Specific loads on the fuselage (except for the central part) are insignificant.

The wing bears loads both from the weight of the plane and wind (from the influence of air flows). In this case, the wing operates in such a way that, during a flight, its upper part (upper panels) is in a compressed state, and the lower part is in a stretched one (when parked on the contrary). Therefore, the upper part of the wing (compressive stresses during flight) is made from high-strength aluminum alloy 7075, which has the maximum specific strength among the commercial aluminum alloys. The lower wing panels are made from alloy 2024, which has higher fatigue strength with lower static strength compared to 7075. In recent years, there has been research focused on the development of next-generation high-strength alloys for manufacturing strength constructions of the aircraft frame, especially the wing (mainly to increase the ductility and fatigue strength limits of the alloys).

The front edges of the wing and tail assembly, heated to prevent icing, are made of heat-resisting aluminum alloy 2618 (Al-Cu-Mg-Fe-Ni alloy system), and on the latest aircraft it is made of alloy 6013 (Al-Mg-Si alloy system) as more technologically advanced if compared to alloy 2618 and capable of operating at temperatures up to $T=180$ ℃.

The control wheels of the aircraft (on the wing and stabilizers), considering the conditions of maximum stiffness (minimum elastic deformation in case of deviation), are made of high-modulus carbon composites.

The stabilizers may be assembled from aluminum alloys, with the approach to choosing the material the same as for the wing. In some aircraft, the stabilizers are a whole integral construction made of carbon composites, which provides a significant weight effect. The polymer matrix composites are also used for wing-fuselage fairing and stabilizers, chassis fairing and so on[4].

As mentioned above, the main criteria for choosing the structural materials, including aluminum alloys, are their specific values of strength, stiffness and fatigue. Nevertheless, corrosion, which exerts in constructions after 10—20 years of aircraft operation, also puts forward its own requirements for designing. Since corrosion damage is shifted in time from the moment of design and manufacture, and the aspects of strength, operation (ease of maintenance) and cost conflict with corrosion resistant requirements, a conscious compromise is necessary when choosing materials at the design stage. This can be confirmed by a classic example of the ratio of corrosion resistant properties and strength characteristics of aluminum alloys, which are subjected to aging hardening (see Fig. 1.5)[5].

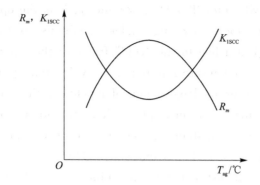

FIGURE 1.5 The ratio of the critical intensity factor for stress corrosion cracking (K_{1SCC}) and tensile strength (R_m) depending on the aging temperature (T_{ag}) of aluminum alloys

As shown in Fig. 1.5, the maximum strength of the alloy corresponds to the minimum corrosion resistance, and in order to ensure the operability of the alloy, it is necessary to underage or overage it, which, as a rule, is done for specific alloys (overaging is more often). When choosing a structural material from the point of view of anti-corrosion properties, it is also necessary to consider the electrochemical potential of the contacting materials, the type and possibility of applying protective coatings, operating conditions, and so on.

The landing gear perceives significant dynamic loads during takeoff and, especially, landing. The landing gear units are made of high-strength titanium alloys (TC18) or high-strength construction steels (30CrMnSiNi). Moreover, titanium alloy in new designs with a long service life is also preferred due to corrosion resistance, as mentioned earlier.

Aircraft systems perform various functions that impose their own requirements on the choice of material for their manufacture. For example, the aircraft's steering wheel, steering column, and pedals cannot be miniature, since the pilot must feel in control. The loads on these components are insignificant; therefore, they are traditionally made of casting magnesium alloys with a minimum specific weight and enough strength for these components. Corrosion resistance for such components is not critical (constant contact with the hands of the pilot). The latest aircraft models have the steering wheel made of polymers using the *rapid prototyping* method.

Pipelines of the aircraft's air system, through which hot air is transmitted (temperature up to $T=350$ °C) from the engine to the front edges of the wing and stabilizers, can be made of thin-walled pipes from corrosion resistant steels, nickel or titanium alloys. Since titanium has the low linear thermal expansion coefficient among metals $\alpha = \Delta L/(L \cdot \Delta T)$, low density $\rho = 4\,500$ kg/m³ and relative high strength $R_m =$ 600 MPa (for Ti-2Al-2.5Zr alloy), it is the optimal material for producing the long pipes operating under conditions of thermal cycles (from -60 ℃ to $+350$ ℃) attached to the "cold" frame of the aircraft. In this case, a minimum number of compensators is required, which remove thermal loads from the airframe, during thermal lengthening of pipelines, the total length of which reaches hundreds of meters in an airplane. The minimum pipe wall thickness is limited by technological capabilities; the weight of the pipe system made of titanium alloy is much less than from steel. The use of nickel alloys in this temperature range is not justified from any point of view (heat resistance, specific gravity). This makes the titanium deicing system economically viable, even considering the high cost of titanium alloys comparing with steels.

Based on the above examples, we can make a conclusion that the choice of materials for parts and units of the aircraft is determined by their design features, operation conditions, external factors, including corrosion and heat, and is a creative process that equally requires knowledge of the properties of materials and aircraft design, operation conditions of its accessory. When choosing a material for a component, it is necessary to consider the whole complex of its properties (physical, chemical, mechanical, technological and operational).

1.4　Role of Materials in Aerospace Engineering

Creation of aircraft, including supersonic and hypersonic aircraft for studying the outer space, determines fundamentally new tasks for materials scientists. Materials for aviation and rocket and space technology should have the following set of properties:

① high specific strength at low and high temperatures;

② stability of properties in a wide range of temperatures during a long period of operation;

③ high resistance to corrosion, penetrating radiation and climatic factors;

④ high resistance to wear, including erosion and fretting;

⑤ high heat resistance and thermal integrity;

⑥ gas impermeability and other properties on which the weight efficiency, reliability and service life depend.

The required level of properties can be achieved both by improving existing materials and creating new materials and technologies for their production intended for operation in extreme conditions[6-8].

Currently, a fundamentally new concept of material has appeared. Previously, the term "material" was understood as a substance of a given chemical composition, which fully determined its physical, mechanical and operational properties. Now the concept of material is presented as a triad: *composition-structure-technology*. New approaches allow to form a given design of a material (configuration or structure) in advance, which is achieved by a certain set of technologies used while producing the material.

Such an early design of the structure and its materialization by choosing the optimal parameters of successive technological processes is the essence of the modern concept of structural engineering of materials. In this regard, it is the powder metallurgy that has enormous potential in the field of designing materials, that is, optimizing their composition and structure, technological regimes, as well as predicting the achievable levels of physical and mechanical properties. With the help of powder metallurgy, it is possible to create materials with unique level of properties, including gradient properties, which generally cannot be achieved using traditional technology of metallurgical production[9-11].

The determining condition for modern aviation itself is the creation of Gas-Turbine Engine (GTE). Creation of such engines becomes possible only after the development of high-temperature alloys. In piston engines (30—40 s of the last century), the operating temperature of materials of the heat zone of the engine was approximately $T = 300$ ℃, and ordinary engine steels were used for their production. To create the first jet engines, materials with operation temperature of 500—600 ℃, which exceeded the creep temperature of steels, were required. New high-temperature high-alloy steels and materials based on nickel were required. Over the past decades, the operation temperature of the blades used in the hot path of the GTE had been raised up to 1 200 ℃ (up to 1 300 ℃ according to some data). The materials for modern engine systems should meet stick requirements for resistance to complex effects of high temperatures, stresses and external aggressive environments. They must have high heat resistance, creep resistance, thermal integrity, thermal stability, low-cycle fatigue, erosion resistance. Today, the threshold has been reached for increasing the operating temperature using traditional methods for parts of the heat path and a raise of up to 10 ℃ is considered as a revolutionary achievement[3-4].

As mentioned above, complex alloys based on nickel are used for manufacturing such parts. These alloys tend to segregate alloying elements in the process of melt crystallization, which complicates possibility to design materials when manufacturing parts therefrom. Using the methods of powder metallurgy eliminates the segregation of alloying elements, prevents formation of coarse inclusions of intermetallic compounds, carbide and carbonitride, which makes it possible to use more doped alloys with a homogeneous structure, as well as improves physical, mechanical and operational properties. In addition, the use of powder metallurgy methods gives more possibilities to improve material properties by creating specific non-equilibrium structures, gradient materials, etc. At the same time, the emergence of many new technologies in powder metallurgy is connected directly with the needs of aviation and astronautics.

New directions in creation of high-temperature materials are dispersion strengthening and mechanical alloying. Due to development of the process of mechanical alloying, it became possible to manufacture dispersion-strengthened high-temperature resisting alloys, which are characterized by the fact that, along with the hardening of the dispersed γ'-phase precipitates (classical hardening phases, ensuring the operation of the alloy at low and medium temperatures), they are additionally hardened by Y_2O_3 dispersed particles, which allow to significantly increase the operating temperatures limit[12-13].

The necessity in high-temperature resistance high-strength materials for engines has determined the emergence of a new class of materials—cermet, which are a combination of metal and ceramic components with a higher melting point and chemical resistance (resistance to oxidation). Such materials are produced by powder metallurgy and have high heat resistance, strength and hardness[13-14].

Powder metallurgy also significantly reduces the cost of products from hard-processing materials, which include heat resistance alloys, due to formation of blanks that are similar in geometry to the finished products and thus reduce the amount of machining and metal waste. During traditional production of parts, the weight of blanks often exceeds the weight of the finished product, which leads to a significant increase of its cost.

The development of jet aviation and space technology has required the development of high-temperature resistant alloys based on refractory metals. In outer space, under the action of vacuum, temperature, radiation and the flow of high-energy particles, the total, especially selective evaporation of metal along grain boundaries sharply increases, which leads to change in alloy composition, impact of erosion and corrosion, and therefore loss of performance properties of space equipment products. The evaporation of volatile components and formation of thin films on the surface of more stable metals causes the appearance of electroplating pairs and, consequently, an increase in corrosion processes. In space conditions, due to the absence of passivation, the compatibility of various metals is of great importance. As a result of surface cleaning, the tribotechnical properties of friction pairs deteriorate dramatically even under conditions of minor loads. When comparing the temperatures at which an equal amount of different metals evaporates from the surface in vacuum, it follows that refractory metals are the most resistant in space, and their properties most closely meet the requirements set for many parts of spacecraft (engines, nozzles, rocket bows, etc.)[15-16].

Powder metallurgy methods make it possible to manufacture structural and functional parts (heat-resistant, refractory, pseudoalloys, alloys with special radio-technical properties, etc.) of refractory metals and compounds at temperatures often much lower than their melting point, preventing chemical interaction with the environment and materials of heating devices.

The main mass of the "flying" materials are light (aluminum, titanium, magnesium, beryllium) and special alloys. Titanium alloys have a unique combination of high strength properties, low density, high corrosion resistance in all-natural environments. The main direction of titanium science today is expansion of the field of application of titanium through the development of new cost-effective technologies that provide a significant reduction in the cost of both the metal itself and finished products made therefrom if compared to existing methods. Powder metallurgy methods can significantly reduce the cost of titanium products. This is achieved due to the possibility of obtaining initially alloyed titanium powders with the subsequent formation of blanks from a mixture of powders corresponding to the composition. In the process of sintering there occurs in-situ alloy formation in the solid phase, which allows to avoid the formation of large-crystalline structure, which is typical for cast titanium alloys. This technology allows to obtain a uniform fine-grained structure, even for large aerospace products.

The use of ultrahigh cooling rates during crystallization allows us to create a new group of aluminum alloys, which are similar to standard aluminum alloys in terms of their structural strength, and it significantly exceeds the latter for other performances (technological plasticity, weldability, etc.). Achieving ultrahigh crystallization rates is carried out by granulating alloys, that is, producing micro ingots by spraying a stream of liquid metal with a gas stream. Subsequently, the granules are compacted and subjected to hot pressure treatment according to standard schemes. The developed granular aluminum alloys are distinguished by a low coefficient of thermal expansion, high heat resistance, an increased Young's modulus, and other unique properties, which makes it possible to use them widely in aviation technology.

Compared to magnesium alloys produced by traditional technology, granular magnesium alloys have higher strength properties and are more economical in production of semi-finished products, which, in combination with low weight, makes them attractive for use in aircraft construction.

Beryllium, which has unique physical and mechanical properties, is a construction material of the 21st century. Good thermal properties of the metal, as well as good oxidation resistance, make beryllium an excellent heat absorber, which is suitable for braking devices of reusable spacecraft and military aircraft. Beryllium is used in rocket nozzles, laser avionics, and the James Webb Space Telescope[17]. It is very promising for hypersonic aircraft[18].

The most important materials used in aircraft include materials for tribotechnical purposes—friction and antifriction materials. It is impossible to imagine any moving mechanism without materials for tribotechnical purposes. With the help of powder metallurgy methods, many materials have been developed for friction units operating under a wide variety of conditions: with lubricant and without it, in air or gas environment and in vacuum, at high speeds and high loads, at elevated temperatures, in aggressive environments etc[19-20].

The control of fast-moving objects with the help of braking devices; energy transfer from the command to the executive parts of a machine through clutches; protection of power units of a machine from overloads due to the use of safety couplings and other similar tasks are associated with the use of friction materials, that is, materials with a high coefficient of friction[21-22]. Sintered friction materials have significant advantages over traditional materials with organic binders. They are efficient at significantly higher temperatures and specific loads. In addition, they have increased wear resistance and stability of properties in the presence of oil or water; they are less affected by arctic or cosmic cold, tropical heat and changes in atmospheric humidity. At the same time, it is possible to replace other types of friction materials with powder ones without changing the design of the friction unit.

Products of aviation and rocket-space technology are characterized by the presence of friction units with unusual and extreme operation conditions that require the stability of their work during a long period of time, considering the effects of external loads and changing external factors (temperature differences over wide limits, humidity, exposure to corrosive environments, dust etc.). For such units, the class of powder antifriction materials, which are developed and operated in relation to certain conditions of the friction unit, is indispensable. Antifriction materials cannot be universal due to the multi-factorial nature of external influences on the friction unit—load, movement speed, type of lubricant, the way applied to the friction surface, the influence of the environment, operating temperatures, etc.

The tactical and technical characteristics of modern aerospace objects of different purposes are largely determined by the characteristics of the radio-technical systems located on board. To protect them from external influences, radio transparent materials are used; these materials should have high radio engineering characteristics, heat resistance, durability over the entire temperature range, impact toughness and low values of thermal conductivity, heat capacity and density[23-24].

The necessity for radio transparent antenna radomes of aircraft is determined by the increment of their speed and maneuverability, the ability to operate in all weather conditions, and increment of requirements for radio-technical characteristics. The main materials for head-mounted antenna radomes of high-speed aircraft are quartz and aluminosilicate ceramics, glass ceramics based on Li_2O-Al_2O_3-SiO_2 and heat resistance glass-ceramic materials (sitalls and glass ceramics) with operating temperatures up to 1 100 ℃[23]. Application of powder technology allows to obtain whole classes of radio transparent glass-ceramic materials with desired properties due to changes in technological parameters[25].

Ceramic materials that are not subject to thermal erosion (Si_3N_4, BN, Al_2O_3, and others) are used for returning spacecraft. The outer surface of the apparatus can be heated to temperatures of 2 000 ℃ and above when the spacecraft entering the Earth atmosphere at a tremendous speed. The operation of ceramic radio transparent materials at high temperatures under aerodynamic braking conditions at hypersonic speeds is provided by ablating the material and reduction of temperature at its surface due to the significant endothermicity of the dissociation and evaporation processes of BN and Si_3N_4. For higher operating temperatures, sintered radio transparent materials based on MgO and Y_2O_3 are recommended[26-28].

Various functional materials are used in rocket-space and aviation technology, ranging from high-density products to highly porous structures used in nozzles, heat shields and heat absorbents, electronic filters, brushes, inertial navigation systems, power systems etc. For comparison, the density of foam aluminum is only 200 kg/m^3 whereas sintered heavy alloys based on tungsten (used to produce antiflatter cargo) are at least 16 500 kg/m^3.

The most important factor for improving the aviation and rocket and space technology for the coming years will obviously be the new materials based on nanotechnology with properties and characteristics that cannot be achieved using current methods. In this regard, these are the methods of powder metallurgy, especially in combination with severe plastic deformation, that are the most rational methods for obtaining nanoscale particles and bulk materials in the nanostructured state[29-32].

Development of aviation transport is determined mainly by the development of engine technology, progress in aerodynamics and creation of new structural materials. Nanotechnology plays important role as methods of developing and manufacturing of completely new structural materials. So, for engines of supersonic airplanes, composite materials with high heat resistance (temperatures up to 1 600 ℃), which are supposed to be developed using nanotechnology, are necessary. Based on nanotechnology, it will be possible to create anti-icing coatings soon, increasing safety of flights by 6—8 times, reducing fuel consumption by tens of percent, improving environmental friendliness and comfort in passenger aircraft[33].

Nanotechnologies have tremendous prospects in the field of auto-electronic emission (to discharge static electricity from an aircraft, control current and fast detonation combustion processes in a hypersonic flow) and provide the ability to absorb electromagnetic radiation by an aircraft airframe[34-35].

Development of space technology is hampered by the high cost of sending cargo into orbit, especially for long-distance flights (for example, outside the solar system). This stimulates the search for methods of reducing the size and mass of spacecraft, as well as improving the efficiency of launch systems. Many of the problems arising therefrom can be solved by using nanostructured materials and devices. Such materials can be particularly useful for manufacturing lightweight, durable and heat-resistant parts of rockets, space stations and research probes for long-range space flights. Moreover, it may be possible, in the conditions of outer space (absence of gravity, high vacuum) to organize research or even the production of such nanostructures and nanosystems that are difficult to obtain on Earth.

In future, the aircraft will be equipped with a multitude of nano-sensors, which will record the information about the air flows during the flight. After it is being processed by the onboard computer, the nanoactivators, acting on the flow, will change the conditions of external aerodynamics in the right direction. This will unprecedentedly increase the efficiency and reliability of aircraft. Special advances in the use of nanotechnology are predicted in their strength. So-called "self-healing constructions" will be created from structured composite materials with interspersed nanoparticles, which will result in the closure of microcracks. Today, the world is on the verge of a grandiose nano-epoch in the aviation and rocket and space industry. The field of possible application of nanomaterials and nanotechnologies in the aerospace industry is truly limitless.

The significant expansion of the range of powder metallurgy products in aviation and rocket and space technology is facilitated by the development of additive manufacturing. Additive technologies are innovative technologies for the direct manufacture and repair of complex and unique parts with high precision from metal powders without using intermediate operations for processing workpieces and special equipment[36-39].

In the aviation industry, the introduction of additive manufacturing is very attractive, since often the efficiency and reliability of a part may depend on its complex spatial form. In addition, the cost of complex parts, especially for important purposes is very high, and the reduction in production time, exclusion of the bulky expensive tooling (casting, forge) is of great interest in the design and development of new products. In addition, the aerospace industry is not characterized by mass production of parts, where additive technologies are difficult to compete with traditional ones.

Additive technologies provide broad prospects for increasing the weighting efficiency of aviation equipment products through more rational design without considering the limitations caused by traditional technologies (casting allowances, transition radius, gradients for pressure treatment, etc.). The transition to the so-called "bionic" design (that is, based on ideas found or borrowed from nature) allows you to change some of the principles of the aircraft design[40].

In addition, since modern alloys (for example, for blades of GTE turbines) have almost exhausted the reserves for improving their properties, additive technologies can become a platform for designing new advanced materials, including gradient ones, that consider the working conditions of a particular part, and are highly adapted to particular use. Development of technologies for manufacturing parts of aviation and rocket and space technology using additive technologies is a transition to the sixth technological period.

The combination of powder metallurgy with other technological areas significantly expands its capabilities, allowing to solve fundamentally new tasks in the field of aviation science of materials. In this regard, the usage of the methods of sever plastic deformation, namely, twin-screw extrusion, equal-channel angular pressing, etc. , are of a great interest. This combination of technologies allows to obtain blank parts with a sub-microcrystalline structure, increased level of mechanical properties and technological plasticity. At the same time, there is no need for the expensive and time-consuming hot processing of ingots and residual porosity, the main flaw of powder products, is eliminated.

Powder metallurgy technologies are constantly being improved and developed. Due to the capabilities of technological methods of powder metallurgy, their flexibility and constant updating, the increment in the potential capabilities of modern materials, as well as expansion of the range of powder materials, which often have unique properties that meet the requirements of the latest developments of aviation and rocket technology, takes place.

Regarding the production of refractory compounds, a relatively new method of Self-propagating High-temperature Synthesis (SHS) is very effective[41-42]. It is a kind of antipode of high-energy effects, since using it does not require large energy demands from the outside (only to start the reaction). In the process of SHS, the isothermal reaction of formation of carbides, borides, silicides, nitrides is used. Under optimal conditions, almost complete conversion of the initial products into the final product of high purity (99.99%) occurs, since nothing is added from outside.

Thus, advances in powder science of materials have largely provided achievements in aircraft and rocket science.

The purpose of this book is to familiarize with basic powder materials, technologies for producing parts therefrom and promising developments in the field of nanomaterials, metal, composite and ceramic materials for aerospace purposes.

Chapter 2 Introduction to Powder Metallurgy

Powder metallurgy—field of science and technology involved in obtaining powder materials, alloys, high-melting compounds and compounds with special properties, as well as materials and products based thereon. As part of science, powder metallurgy is an integral part of science of materials—science of structure and properties of technical materials. As technology, powder metallurgy creates and develops the technology of production of sintered materials and products, which excludes melting of the base material.

Traditional technology of powder metallurgy includes the following sequence of operations:

① Obtaining the powder—we distinguish physical and chemical as well as mechanical methods of production of powders. Physical and chemical methods include technological processes of production of powders associated with physical and chemical transformations of the raw material (chemical reduction, electrolysis of solutions or melts, dissociation of carbonyls, thermal-diffusion saturation, sublimation and condensation, intercrystalline corrosion, electrical erosion). *Mechanical methods* ensure the transformation of the raw material into powder without a noticeable change in the chemical composition (crushing and grinding, spraying, granulation, blade processing—cutting). The main methods for producing powders and the powders produced using these methods are given in Table 2.1[43].

TABLE 2.1 Main methods of obtaining the powders

Method of obtaining the powders	Characteristics of the method	Obtained powders
Physical and chemical methods		
Chemical reduction of oxides and other solid metal compounds	The most common and economic method. Reduction agents—gases (hydrogen, convertible natural gas), solid carbon (coke, colloidal carbon), metals (Na, Ca, Mg). Raw materials—ores, concentrates, metallurgical production wastes, as well as various chemical compounds of metals	Fe, Cu, Ni, Co, W, Mo, Ta, Nb, Zr, Ti and other metals and their alloys

(continued)

Method of obtaining the powders	Characteristics of the method	Obtained powders
Vapour phase reduction	Reduction of gaseous compounds with hydrogen in a fluidized bed or plasma	Ta, Nb, Ti, Th, Zr, V, U
Electrolytic deposition of aqueous solutions or fused salts	The deposition on the cathode under the influence of electric current from aqueous solutions or molten salts of pure powders. Peculiarity—high cost of powders due to the high cost of electricity and low productivity of electrolytes	Cu, Ni, Pb, Fe, Ag, Zn—from aqueous solutions; Ta, Nb, Fe, Th, U, Zr, Be, Ti—from melts
Dissociation of carbonyls	Dissociation by heating of metal compounds of the $Me_x(CO)_y$ type. High-quality dispersed powders (expensive) are obtained	Fe, Ni, Co, Cr, Mo, W, Mn
Condensation of gaseous metal	Metal evaporation and steam condensation on a cold surface. Finely dispersed powders with a high content of oxides are obtained	Zn, Mg, Cd etc. Metals with low T_{ev}
Intercrystalline corrosion	Destruction in a compact (cast) material of intercrystalline interlayers using chemical etchant	Corrosion resistant and nickel chrome steels
Method of electrical erosion	Sputtering of metal from the end of the electrode in the layer of electrically insulating material	Any metals and alloys
Mechanical methods		
Crushing and grinding	Grinding of shavings, scraps and compact materials in various crushers and mills	Fe, Cu, Mn, Cr, Al, brass, bronze, steel
Atomization	Dispersion of a jet of molten metal by mechanical means (by rotating blades, centrifugal forces) or by the jet of the energy carrier (gas or liquid)	Al, Pb, Mn, Cr, Zn, Ni, Fe, Ti, Sn, bronze, brass, cast iron, steel
Cutting	Machine processing of compact (cast) metals	Steel, brass, bronze, cast iron

② Mixing of powders (sometimes with an appropriate lubricant, plasticizer or binder).

③ Compaction of powders (or their mixtures—batch mixture) in molds (or without them) to obtain a *compact* (press forming, briquette) with sufficient cohesive strength between the powder particles, providing safely operation with the compact. Such compact is called "Green", hence the terms "Green Density" and "Green Strength".

④ Sintering—heating of the compact (usually in a protective environment) at temperatures below the melting point of the main component so that the powder particles are welded together to ensure sufficient strength for the intended use of the materials (sintered part). Sometimes, when sintering, non-main components melt, then the process is called "Liquid Phase Sintering", which allows to intensify the processes of consolidation of powder particles. Based on the applied physical and chemical processes the liquid phase sintering is divided into sintering with disappearing liquid phase (formation of compounds) and with available liquid phase till the end of sintering (no interaction between the components of the mixture). The processes of pressing and sintering are sometimes combined, producing the processes of "Hot Pressing" and "Pressure Sintering". Sometimes (in particular for highly porous materials of filtering components), the pressing operation is not used, and the powder is sintered in a special form. This process is called loose sintering[44].

⑤ In most cases, the sintered part is subjected to additional processing in order to ensure dimensional accuracy and necessary operational properties—calibration (repressing), machining, electrodepositing and paint application, etc. The maximum effect from the technology of powder metallurgy is obtained in case of obtaining the finished part without machining (no machining).

The powder metallurgy technologies solve the following two major problems in the production of parts, which also relate to the aerospace applications:

① Production of materials and products with a special composition, structure, and therefore properties that cannot be obtained by other methods.

② Production of materials and products with usual composition, structure and properties, but with much more favorable economic benefits of their production.

2.1 The Properties of Powders

2.1.1 Physical Properties

Metal powder—a set of particles of a metal, alloy or metal-like compound with dimensions up to 1 millimeter, which are in mutual contact and not interconnected.

Particle size distribution—the size range of particle between their minimum and maximum values. The *granulometric composition* of the powder or size distribution of particles gives information on the particle number with specific size range in fractions (%) in relation to their total number. The granulometric composition is presented in the form of graphs, tables and histograms. The dependence of the particle content to their size can be discrete and continuous. The granulometric composition is determined by various methods of analysis—sieve analysis, sedimentation, quantitative image analysis, laser diffraction and others[45-46].

Particle shape significantly affects the technological properties of the powder, as well as the strength, density and homogeneity of properties of products therefrom. Depending on the methods of obtaining and processing the powder, the particle shape may be spherical, drop-shaped, fibrous, fragmentation, spongy, dish-shaped, scaly, dendritic, irregular. Spherical, drop-shaped and dish-shaped forms are typical for spraying powders; fragmentation and scaly forms are typical for mechanical grinding; spongy and irregular forms are for restoration; dendritic is for electrolysis. The shape of the particles is characterized by optical and electron microscopy.

Pycnometric density of a powder particle is always slightly different from the theoretical (X-ray) density of a powder substance. This is because powder particles have impurities, closed porosity, crystal-lattice defects and so on. Pycnometric density is determined by the method of pycnometry, which consists in weighing a substance of a certain volume[47].

Microhardness of particles of the powder characterizes its ability to deform, which is important for assessing the possible behavior of the powder during pressing. The magnitude of the microhardness largely depends on the distortion of the crystal lattice distortion due to residual stresses and the presence of various impurities in the powder particles.

Specific surface of a powder is the total surface of all particles that make up a unit of mass or volume; determined by the dispersion, shape and surface state of its particles. The activity of physical and chemical processes associated with the particle properties, their compaction, sintering, interaction with environment etc., depends on the specific surface magnitude. The determination of the specific surface is an important point in the practice of powder metallurgy. It is determined by two main methods—the adsorption method and the method based on permeability, i.e., air filtration through a layer of powder[48]. The adsorption-structure behavior of the powder is determined on the basis of the adsorption isotherms of gas adsorption (nitrogen) obtained by the Brunauer-Emmett-Teller (BET) method[49] (the theory of gas molecular adsorption proposed by S. Brunauer, P. H. Emmett and E. Teller):

- specific surface S_{BET} (m^2/g).
- total pore volume V_p (mL/g).
- adsorption potential A (J/g), which can be expressed as

$$A = \frac{RTV_p}{p/p_s} \tag{2.1}$$

where p/p_s—relative pressure;

p_s—saturation pressure;

R—universal gas constant;

T—temperature.

- specific adsorption potential A^* (J/m^2), which can be expressed as

$$A^* = \frac{A}{S_{BET}} \tag{2.2}$$

- latent heat of nitrogen adsorption E, which simultaneously takes into account the activity of the surface centers and their number:

$$E = RTV_m \ln C_{BET} \tag{2.3}$$

where V_m—volume of nitrogen monolayer on sample surface;

C_{BET}—the adsorption coefficient[50].

Morphometric parameters of powder need relatively new method for detecting powder characteristics. The method is based on processing of data obtained on the DiaInspect device. DiaInspect-OSM produced by Vollstadt Diamant GmbH allows to automatically determine more than 20 measured characteristics of the sample powders. The main morphometric characteristics of the powder are (see Fig. 2.1)

- elongation

$$F_{eret} = \frac{F_{max}}{F_{min}} \tag{2.4}$$

where F_{max} and F_{min}—maximum and minimum distance between two parallel tangents to the grain contour.

- circularity

$$C_u = \frac{P_u^2}{4\pi A_0} \tag{2.5}$$

where P_u—actual grain perimeter;

A_0—grain projection area.

- convexity

$$C_B = \frac{P_B^2}{4\pi A_0} \tag{2.6}$$

where P_B—perimeter of the convex image of the grain.

- ellipticity

$$E_l = \sqrt{\frac{J_{max}}{J_{min}}} \tag{2.7}$$

where J_{max} and J_{min}—invariant complexes of axial moments of inertia J_x and J_y relative to the axis x, y of rectangular coordinate system and centrifugal moment J_{xy}.

Invariant complexes are expressed through the indicated moments by dependencies:

$$J_{max} = 0.5\left[J_x + J_y + \sqrt{(J_x - J_y)^2 + 4J_{xy}^2}\right] \quad (2.8)$$

$$J_{min} = 0.5\left[J_x - J_y + \sqrt{(J_x - J_y)^2 + 4J_{xy}^2}\right] \quad (2.9)$$

- roughness R_g which characterizes the smoothness of the projection contour and is determined by

$$R_g = \frac{P_u}{P_B} \quad (2.10)$$

- specific perimeter P_o (μm^{-1}) which characterizes the smoothness of the projection contour and represents the ratio of the actual perimeter of the projection of the grain to its area, i.e.

$$P_o = \frac{P_u}{A_0} \quad (2.11)$$

- average grain size d_m (μm) which is calculated by

$$d_m = \frac{F_{max} + F_{min}}{2} \quad (2.12)$$

- equivalent grain size

$$d_e = 2\sqrt{\frac{A_0}{\pi}} \quad (2.13)$$

The above listed characteristics of the powders describe the grain size (elongation F_{eret}, F_{max} or F_{min}, average grain size d_m, equivalent grain size d_e), their shape (circularity C_u, convexity C_B, ellipticity E_1) and surface topography (specific perimeter P_o, roughness R_g), as shown in Fig. 2.1[51].

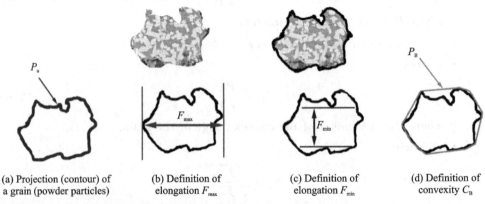

(a) Projection (contour) of a grain (powder particles)
(b) Definition of elongation F_{max}
(c) Definition of elongation F_{min}
(d) Definition of convexity C_B

FIGURE 2.1 Determination of morphological properties of powder

2.1.2 Chemical Properties

Chemical composition largely depends on the chemical composition of the initial materials for their preparation and the method of producing powders. Depending on the concentration of the base metal, the powders are usually divided to: highly pure, pure and technically pure, the concentration of the main component therein is, 99.99%, 99.5% and 99.98%, respectively. The maximum concentration of impurities in powders is determined by their permissible concentration in the finished product.

Metal powders contain a significant amount of gases (O_2, H_2, N_2) both adsorbed on the surface and in the form of interstitial solid solutions or chemical compounds—oxides, hydrides, nitrides. The presence of a significant amount of gases increases the fragility of powders, makes molding difficult, and their intense release during sintering can lead to distortion of products.

It should be noted that the purity of powder is a critical characteristic. Allowed levels of impurities largely depend on the nature and state of the combination of substances. For example, the presence of chemically bound carbon in iron contributes to the hardening and embrittlement of the matrix, and therefore requires increased pressure during pressing and increases tool wear. However, free carbon is often useful because it acts as a solid lubricant during pressing. Particles of most metall powders are coated with thin oxide films, which are destroyed during pressing forming clean, active metallic surfaces that are easily joined by cold welding during deformation, providing "green strength". Their final recovery during sintering in a controlled reducing environment is an important condition for activating sintering processes, ensuring metal bonding and achieving maximum strength. On the other hand, the presence of oxide films or inclusions of oxide particles (SiO_2, Al_2O_3) is a serious negative factor since it is an abrasive that wears the press tools. Such oxides are not reduced by hydrogen during sintering and their presence adversely affects the mechanical properties of the sintered products, especially the toughness of the material.

Pyrophoricity—the ability of powders to spontaneous combustion in contact with air depends on the nature of the metal, the degree of dispersion and shape of the powder particles, the state of their surface, purity, etc.[52]

Toxicity—the ability of powders to cause biologically harmful effect on the human body. Almost all metal powders have a harmful effect on the human body if contact with it, although in a compact state most metals are harmless. The ingestion of powder into the human body is possible through the respiratory system (most often) or the digestive organs[52].

2.1.3 Processing Properties

Apparent density is the mass per unit volume of loose packed powders. The apparent density is a function of the physical density of the powder material, the size and shape of particles, the fractional composition of powder, the packing density of particles and the state of their surface. The apparent density is used for determining mold sizes. The apparent density is determined by the method of volumetry as the ratio of the mass of apparent powders to volume, which is defined as $\gamma_{appar} = \dfrac{m}{v}$ [52-53].

Tapped density, *tapped volume*—powder density after mechanically tapping or vibrating the metal powder under specific conditions, depends on the same powder parameters as apparent density. The fractional composition of the powder has a special influence; the presence of small and large particles increases the tapped density. As a rule, the density of the tapped density exceeds 20%—50% of the apparent density. The ratio of apparent density to the tap density gives a qualitative idea of the fluidity of the powder. The reciprocal of the tap density is called the tap volume, which corresponds to the volume occupied by the unit mass of the tapped powder[53].

Angle of repose—the angle between the horizontal and the cone forming free-apparent powder, depends on the degree of dispersion of the powder, the shape of the particles and the coefficient of external friction between them. The angle of repose characterizes the equilibrium state of the powder in the absence of exposure to external forces[54].

Flowability—the ability of the powder to move under the action of gravity, depends on the density of the material, the fractional composition of the powder, the state of the surface and the shape of the particles, the degree of oxidation, the presence of adsorbed moisture and other factors. The flowability of the powder is of practical importance because it determines the time and completeness (quality) of filling the molds, especially with automatic pressing[55].

Mouldability—the ability of the powder to form a body with given shape, size and density under influence. Mouldability is a complex characteristic of powders, which is determined by their compaction and formability[56].

Compressibility—the ability of powders to change the density at the applied external pressure. To determine the characteristics of compressibility the charts of pressing are built. They are graphs of dependence of density of compacts on the pressing pressure. The compressibility is affected by the whole complex of powder properties and pressing conditions[57].

Formability—the ability of the powder to maintain the shape given to it as a result of pressing in a given porosity range, depends on the hardness, shape, size and surface condition of the powder particles. In practice, formability determines the pressure range required for pressing the powder samples.

The choice of the characteristics of the powder is based, as a rule, on a compromise, since many factors contradict each other. For instance, an increase in the irregularity and porosity of powder particles, that is, a decrease in its apparent density, increases compressibility during pressing and, thus, the degree of cold welding, which in turn increases the green strength of the compact. Such an increase in the contacting surfaces also contributes to more efficient sintering. However, higher compaction may require higher pressing pressure, and therefore more powerful and expensive presses, more durable and massive tooling.

The ease and efficiency of powder packaging largely depends on the size distribution of particles so that the voids formed between large particles can be filled with small particles. Fine particles form fine pores that easily disappear during sintering. However, an excess of fine fractions reduces the rheological properties of the powder.

2.2 Producing Sintered Products

2.2.1 Powder Preparation

Annealing of powders are used to increase their plasticity and compressibility mainly due to the reduction of oxides and removal of internal stresses in particles. Heating is carried out in a reducing environment (sometimes in an inert or vacuum) at a temperature of $(0.4—0.6)T_{mel}$ (melting point) of the powder material or the least refractory main component of the powder mixture.

Classification—separation of powders according to the particle size into fractions, which are then used either directly for molding, or for making a mixture containing the required percentage of particles of the desired size. Depending on the devices on which the classification is made, or working environments, there are vibration, air, liquid and screen classification.

Mixing—preparation of a homogeneous mechanical mixture from metal powders of different chemical and granulometric composition or a mixture of metallic and non-metallic powders. The homogeneity of the mixture is characterized by the degree of mixing and depends on the physical density, apparent density, duration of mixing, the size and shape of the particles, the mixing environment, the type of mixer, etc.

The actual duration of mixing ranges from dozens of minutes to several hours. Excessive mixing should be avoided, as it may result in increase of apparent density and deterioration in rheological properties of the mixture (good fluidity is required to load the powder into the matrix, especially with automatic pressing), and the green strength of the compact.

While mixing, various lubricants (stearic acid, metal stearates and other organic compounds of a wax nature) can be introduced into the powder mixtures in order to reduce friction between the powder mass and the working surfaces of the molds, along which the powder slides during compaction. This ensures a uniform density over the entire height of the compact and facilitates the extraction of the compact from the matrix, which minimizes the tendency of crack formation. The choice of lubricants should be approached with caution in order to avoid contamination of the finished products by the decomposition products of lubricant during sintering. For some powders (e. g., titanium), lubricants are not used due to their high chemical activity and chemical interaction with the lubricant during further sintering.

2.2.2 Powder Forming

Pressing—forming powder in the mold under the influence of external pressure. As a result of the deformation, the initial volume of the powder body is reduced and a compact is formed with a given shape, size and properties. Pressing is a critical operation in the process, since the final shape and mechanical properties, their uniformity in the product, depend on the uniform level and density of the pressed compact.

During pressing due to seizing and friction of particles against the walls of the matrix and the surface of punches, the pressure distribution in the powder is uneven, the pressure on the walls of the matrix is substantially less than the pressure in the direction of pressing. The compaction process during pressing consists of two stages, proceeding sequentially, but with certain overlap: first, the particles are transferred with mutual displacement, and then elastic and plastic deformation or crushing of particles.

The presence of external friction of the pressed powder against the walls of the molds determines the force required for the compact after it has been pressed which is called the *repressing pressure*. Typically, the repressing pressure is (0. 20—

$0.35)P_{pr}$ [58-59].

The phenomenon of increment of the size of the compact when removing the pressing pressure, as well as during its repressing out of the mold, is called the *elastic after-effect*. The magnitude of the elastic after-effect depends on the characteristics of the powder to be pressed, pressing pressure, the presence and amount of lubricant, and other factors[59-60].

The main constructional components of the molds are the matrix in which the powder is poured before pressing and the punches used to apply the pressure during pressing. Rods (needles) are used for forming thorough holes, which are placed in the matrix before filling it with powder. For pressing products of complex shape, upper and/or lower composite punches with several moving parts are used, whose number corresponds to the number of transitions (size changes) along the height of the product[59].

There is no need to emphasize the importance of the exact dimensions and high quality of the mold surface, since the main advantage of powder metallurgy is the ability to produce parts with exact size and with a finished surface that does not require further machining (no machining). Another important indicator is the service life of the matrix, which depends not only on the material to be pressed, the required compact density, lubrication, which minimizes the wear of the matrix, but also on the complexity of the equipment and the skill of the operator. In steel die, the compact removal cycles are 200 thousand, and for hard-alloyed die it is up to million or more.

With sufficiently high temperature $(0.5-0.8)T_{mel}$, the processes of pressing and sintering can be combined, and this process is called *hot pressing*. Hot pressing is pressing of metal powder or powder forming at a temperature higher than the recrystallization temperature of the main component. The processes during hot pressing proceed much more intensively than during cold pressing and sintering, which makes it possible to obtain virtually nonporous products. Hot pressing is used to process difficult-to-press and difficult-to-sinter powders of refractory compounds. The use of hot pressing is limited by the high cost: special heat-resistant die and controlled atmosphere are required, and the productivity is low.

Pressing of metal powder at temperatures ranging from 0.1 of melting point to the temperature at which the main component begins to recrystallize is called *warm pressing*[61].

Isostatic pressing is forming molds of compacts in shells with comprehensive compression. At *Cold Isostatic Pressing (CIP)*, the powder is placed in an elastic shell, for example rubber, polyurethane or others, which is evacuated, and then immersed in liquid (usually water) and compressed at high pressure in hydrostats. In this way, a comprehensive compressibility of the powder takes place. Since this does not require lubrication, a high and uniform density of the compact is achieved. This process allows to overcome many restrictions in geometry of products (long-length thin-walled cylinders, parts with undercuts). The process is automated and, in some cases, the productivity is commensurable with pressing in press molds.

Hot Isostatic Pressing (*HIP*) is isostatic molding to obtain the maximum density of the blank at temperatures above T_{rec}. At HIP, an inert gas (Ar) is used instead of a liquid compressive medium, and a metal container is used instead of an organic elastomer shell. The powder is poured into a container, consolidated by vibration, evacuated and filled up. After that, the container is placed in HIP machine, where the HIP process is performed. The advantages of the HIP are: the possibility of obtaining products of complex shape (turbine blades), regularity of the chemical composition of the material and structure; manufacturing of parts from hard-to-press and hard-to-sinter powders; ensuring uniform and high density of the product. HIP is used for pressing materials that are not compacted in a cold state even at high pressures, when they want to combine the processes of pressing and sintering or to obtain a material with special properties and structure[62-63].

An improved version of the traditional HIP is *Net-shape powder HIP*, which provides improved quality of produced parts by improving the shape and special equipment for forming the part from the powder through the use of computer simulation, and also increasing modern ultrasound techniques. This technology allows to obtain large-sized products (including thin-walled, with a wall thickness of 0.4—1.0 mm) having a fine-grained structure. Net-shape powder HIP parts made of titanium alloy have a strength of $R_m = 1\,060$ MPa, whereas $R_m = 900$ MPa with the injection method.

Consolidation by Atmospheric Pressure (*CAP*) is one of the methods of quasistatic pressing. The technology includes the manufacture of a glass shell; activation of the surface of the particles of fine powder by introducing boric acid or a solution of boric acid in methanol, which ensures the destruction of the oxide film on the surface of metal particles; filling the capsule with powder and its evacuation with subsequent sintering. In the process of holding at the sintering temperature, the shell softens and is subjected to slow compression under the action of the difference between the external (atmospheric) and internal pressure (vacuum). Under the conditions of quasi-isostatic pressing, low-porous powder products are formed (with residual porosity of 1%—2%). Subsequent hot processing with a degree of deformation of more than 30% leads to complete dense product. CAP technologies are utilized to make products from superalloys.

Ceracon-process (ceramic consolidation) is a process of quasi-isostatic consolidation of powders under high-pressure, which uses granular ceramic medium as pressure transfer instead of the gas used in HIP. The blanks of compacted material are placed in a granular medium to which uniaxial pressure is applied. At the same time, the particles of the granulated medium easily "flow around" the blank and redistribute the applied pressure in a quasi-isostatic manner. The ceracon process is controlled by introducing different ratios between the shear and normal stress components due to the design of the matrix and the choice of pressing medium. The ceracon process is used to consolidate hard to flow powders, such as Ti_3Al and TiAl. The deformation of such powders is difficult due to the fragile of intermetallic compounds and the presence of stable Al_2O_3 on the surface of the particles. Using the ceracon process at a temperature of 1 200 ℃ under pressure of 1.24 GPa for 45 s, non-porous TiAl is obtained[64].

Vibration pressing consists in compaction of powders in a mold by repeated exposure to a pulsed force of relatively small amplitude. The positive effect of vibration on the compaction process is associated on the destruction of the initial interparticle bonds (in particular, the arches formed by particles) and the improvement of the mutual mobility of particles. As a result of the imposition of vibration on the pressing process, a high density of particle placement is achieved (95% or more of the theoretically possible for a given granulometric particle composition of the powder). Vibration is most effective for compacted powders, which are a predetermined set of fractions of particles with various sizes.

Impulse forming—molding at which compressibility is carried out by shock waves in the time interval not exceeding 1 s. In the case of pulsed molding of the powder in a state of shock, the loading speed is 50—100 m/s. When this occurs, the intensive plastic flow of the compression material and the strain hardening, and destruction of the particles associated with it, occur. High loading rates lead to a decrease in the lateral pressure coefficient, as well as external and interparticle friction, which greatly facilitates the process of increasing the density of the pressed powder body[59]. For impulse loading, explosives, compressed and combustible gases, electro-hydraulic discharge, pulsed magnetic field and so on can be used.

The explosive pressing is used for both loosely apparent powder poured into an elastic shell, and blanks, obtained by preliminary pressing or pressing and sintering. In the first case, compacts with a uniform density of 90% can be obtained, and in the second case, a theoretical density of 99.3%—99.9% can be obtained.

Explosive pressing provides obtaining samples with higher strength than with conventional static pressing with the same compact density. This is due to the increase in the total area of metal contacts, as well as the active destruction of surface oxide films during explosive pressing, which also contributes to the improvement in surface contact.

Extrusion molding—formation of blanks from a mixture of powders by pushing it through a die of a smaller cross-sectional area. The method allows to obtain lengthy blanks (rods, pipes, profiles) with a uniform density of difficult-to-extrude powders of refractory metals and compounds. The quality of the samples depends on the plasticizer, which should provide sufficient viscosity of the mixture and the "green" strength of the blanks before sintering, and also be completely removed during sintering without interacting with the powder. Alcoholic solutions of bakelite, starch paste, polystyrene solutions and others are used as plasticizers.

Slip casting—forming products from slips (concentrated suspensions of powders). The main methods of slip casting are:

① casting into porous molds—the slip is poured into the mold (most often plaster); standing up till the formation of the blank, which is molded due to the action of capillary suction forces, causing the flow of fluid to the walls of the form.

② casting from thermoplastic slips (hot casting)—heating the thermoplastic slurry to its pour point, forming and subsequent cooling to solidify.

③ forming by electrophoretic method—the gradual build-up of a layer of slip particles, moving under the influence of an electric field to the electrode-form and deposited thereon.

Slip casting can produce products of complex shape, miniature and large-size, hollow, with a uniform wall thickness, with a uniform volume density, a high surface accuracy and accurate dimensions[59].

Injection molding—forming of blanks of complex configurations from plasticized powder mixtures. The method is that a mixture of a thermoplastic carrier (a composition containing wax, liquid lubricants, resins) and metal or ceramic powder is injected under pressure into a mold, and the blank is subjected to preliminary heat treatment to remove the components of the carrier, after which it is sintering. The result is a product with a density of about 99% of theoretical density, accurate dimensions and high surface accuracy[65].

Powder rolling—continuous forming of blanks of powder by rolling. The main difference between rolling of powders and rolling of compact material is significant change in their properties—the transition from apparent to compact state. Rolling methods differ by the location of the axis of the mill rolls (vertical, horizontal or inclined rolling), the condition of the material being processed (hot, warm and cold rolling), powder feed (gravity or forced) and the form of rolling (sheet, profile).

Interesting results are shown by asymmetric rolling—rolling performed by using different-sized rolls, which lead to their different linear speed and the occurrence of shear forces in the deformation zone, which makes it possible to obtain more dense and durable blanks compared with conventional rolling[66].

2.2.3 Sintering

Sintering is the thermal treatment of powder and/or a mixture of powders and/or fibers and/or moldings at temperatures below T_{mel} of the main component, carried out in order to consolidate and provide a certain set of mechanical and physical and chemical properties. Sintering results in transformation of fragile powder compact into a strong sintered product with properties approaching (and sometimes exceeding) the properties of the cast and deformed metal, which is caused by the change in the interparticle contact. So, if at compact or apparent powder the particles are mechanically interconnected, then in the sintered product the substance belonging to several particles are formed at once.

Liquid Phase Sintering (*LPS*)—sintering of multicomponent systems with the formation of liquid phase. Specialists distinguish sintering with the liquid phase, which is present until the end of the isothermal holding during heating, and sintering with the liquid phase, which disappears soon after its appearance despite the continuing heating. The liquid phase is formed when the low-melting component is melted or the low-melting component (eutectic) melts and disappears as a result of alloy formation. It facilitates the development of adhesion forces between the powder particles, provided they are wetted. In case of poor wettability, the liquid phase inhibits sintering, preventing compaction[59,67]. Liquid phase sintering is used to activate shrinkage and obtain products with low or even zero porosity and high properties. Examples of powder systems subjected to liquid phase sintering are Fe-Cu, Cu-Sn, WC-Co, Cu-Pb and so on[59].

Sintering atmosphere—environments that intensify shrinkage or protect the blank from chemical interaction with air during sintering. The use of protective media is arisen from the necessity to protect the sintered materials from oxidation during heat treatment, as well as for removal of oxide films available on the surface of the particles.

In practice, dehydrated hydrogen, dissociated ammonia (75% H_2 and 25% N_2) and protective gases obtained by incomplete burning of hydrocarbons are used. In many cases (for example, when sintering products made from alloys of iron and copper), except for cases when reaction with nitrogen is not allowed, dissociated ammonia can replace hydrogen. The cost of ammonia is about three times lower.

The most common (due to low cost) environment is obtained by incomplete burning of hydrocarbons. At the same time, because of the variety of the air－gas ratio, a wide range of compositions is obtained. Hydrocarbons (methane, propane, butane) reacted with a small amount of air can contain $30\%-50\% H_2$, $18\%-25\% CO$, $28\%-48\% N_2$. Since such endothermic reactions occur when heated, the resulting medium is called *endogas*. If the hydrocarbon is burned with a large amount of air, the resulting medium can contain up to $5\% H_2$ and a large percentage of fixed nitrogen. This reaction is exothermic and proceeds without heating from outside, and the resulting medium is called *exogas*. It is the cheapest medium, but its reduction potential is low, so the removal of oxides from powder compacts is ineffective, which negatively affects the strength of the sintered products.

Since the sintering medium must have a certain carbon potential, during sintering, for example, of carbon steel, it is necessary to prevent both decarburization and carburization. The *carbon potential* is a qualitative characteristic of the reactive ability of a gas tending to the "gas－solid" system to provide an equilibrium carbon content[68].

Vacuum can be considered as a special case of a controlled environment. During *vacuum sintering* the gas inside the compact is easily removed and easily oxidized metals (Al, Be, Nb, Ti, Ta) whose oxides are not reduced, even in hydrogen, can be sintered. However, vacuum sintering is quite expensive and not very productive. Nevertheless, during vacuum sintering, it is possible to combine such processes as decomposition and removal of one (several) component(s) in the initial components of a mixture of powders, which allow to significantly intensify the processes of consolidation and synthesis of alloys during sintering. Modern developments in the field of vacuum equipment allow to perform vacuum sintering in a continuous mode.

As a result of sintering, the sizes of the heated powder body change, although in some cases the overall dimensions may remain unchanged (non-shrinkable materials). Specialists distinguish *volumetric shrinkage* and *linear shrinkage*, representing the ratio of the size difference between the initial (before sintering) and final (after sintering) parameters to the initial parameter.

During sintering, the size of the compacts is also possible to increase (increase in their volume). The main reasons for the growth of single and multicomponent systems are the removal of internal stresses arising during pressing, the presence of dissolved oxygen in material, phase transformations and the evolution of gases. In addition, the growth during sintering of a mixture of powders of mutually soluble metals at early stages of sintering can be caused by the inequality of the partial coefficients of heterodiffusion. The growth phenomenon can also be observed during liquid phase sintering.

Activated sintering means conscious application of additional treatments, which, by enhancing the structural activity, affect the underlying sintering kinetics process. The purpose of such treatments is to reduce the time to achieve the given density and strength, eliminate irregularities in the volume of the sintered body and consequently affect the properties, i. e., improving ductility and impact strength or reducing the sintering temperature, and for multi-component systems to balance concentration[69]. During activated sintering under the influence of additional factors that increase the diffusion activity of the sintered material, the acceleration of the sintering process and the intensification of changes in the properties of the material occur regardless of pressure and temperature. Sintering activation methods can be divided into two groups:

① chemical, based on the use of chemical reactions of reduction-oxidation, dissociation of oxides, halides, hydrides;

② physical, including a cyclic change in sintering temperature, influence of ultrasound and various types of radiation, sintering in a magnetic field, and so on.

Activating of sintering by changing the composition of the gaseous medium is one of the simplest and most effective ways to harden products and replace the oxide contact with a metallic one. The activated medium can also favorably influence on the sintering process due to removal of impurities and refinement of the sintered material[59].

2.3 Additional Processes After Sintering

As noted earlier, the main advantage of powder metallurgy is the possibility to obtain finished products out of powder. However, in some cases and for some materials it is impossible. In addition, the operating conditions of specific parts impose additional requirements to their properties, which bring to necessity in additional processing of sintered products. Additional processing is most often carried out in order to protect the surface from corrosion or wear, increase physical and mechanical properties, bring products to final exact dimensions and shapes, apply decorative coating.

Repressing—additional pressing of sintered blanks. After repressing, as a rule, annealing or repeated sintering follows. Repressing is done in order to increase the density of the sintered blanks and increase their physical and mechanical properties[58].

Hot forging—free forging or forging in closed dies at high temperatures in order to obtain dense products from difficult-to-form metals and alloys (refractory metals, dispersion-hardened alloys), ensuring an optimal structure.

Calibration—a kind of pressure treatment that allows to obtain a product with exact dimensions and/or shape due to plastic deformation of the surface layers. In powder metallurgy, calibration is almost always a finishing operation that combines optimization of both size and shape. Calibration is performed in special sizing dies, sometimes after pre-oiling of blanks. Calibration is usually carried out at a pressure of $(0.1-0.25)P_{pr}$.

Infiltration—operation of filling the pores of a sintered or compacted material with a molten metal or alloy having a melting point which is lower than the impregnated material. This is the easiest way to get a compact non-porous material or a consolidated component of a composite material. At the same time, it is possible to increase the service properties or give new properties to the composite. The condition for infiltration is wetting the impregnated material with the liquid phase—the smaller the wetting angle the higher the temperature, then the higher is the speed and completeness of the infiltration.

Impregnation—operation of filling the pores of blanks with liquid substance (lubricants, resins and other media, except for melts) to give special properties to the products. The use of plastics for impregnation contributes to the improvement of mechanical properties and closure of thorough (surface) pores.

Heat treatment is used to increase the mechanical properties, and sometimes to increase the wear and corrosion resistance of powder materials. Sintered materials can be hardened as the same with the cast, but the existence of porosity must be considered.

Thermochemical treatment is heat treatment in a specially selected active saturating medium, which ensures obtaining a predetermined controlled change in the chemical composition of the material surface with a depth gradient. There are many varieties of chemical heat treatment (carbonization, nitriding, carbonitriding, aluminizing, chromizing, chrome aluminizing, boronization, etc.). The most common in the technology are carbonization (cementation), nitriding and carbonitriding. As a rule, thermochemical treatment is performed in order to improve the tribological properties of parts of friction units, sometimes to increase corrosion resistance. Thermochemical treatment of powder products has a high kinetics compared with cast parts due to residual porosity and fine-grained structure.

Steam treatment of powder parts consists in keeping high-pressure at sharp steam at temperature of about 500 ℃. With this treatment there occurs formation of a dense layer of magnetite (iron oxide) on all open surfaces (including in open pores), which leads to improvement of properties—corrosion resistance, compressive strength, wear resistance.

Using the *electroplating method*, you can apply various protective coatings (Sn, Cu, Ni, Cd, Zn, Cr) to sintered parts. In this case, low-density parts must be sealed, for example, by resin impregnation, before coating process is applied to prevent electrolyte from entering the pores and further corrosion. Parts with porosity of less than 5% do not require special measures for sealing the pores.

Physical Vapor Deposition (*PVD*) allows to apply a range of protective (wear- and corrosion-resistant) metallic and non-metallic (nitride, carbide) coatings without further processing of pores.

Electron beam coating—the method that allows to apply protective current-conducting and non-current-conducting coatings, including high-temperature ceramic coatings, of considerable thickness.

Because of availability of restrictions in the geometry of the parts obtained by pressing and sintering, it is necessary to apply further *mechanical treatment* to obtain, for example, holes arranged at an angle to the direction of pressing. For sintered parts, traditional types of mechanical treatment are used—turning, drilling, milling, grinding, etc. It should be noted that the presence of pores leads to increased wear of tools compared with cast materials, and that mechanical treatment immediately increases the cost of powder products and, therefore, mechanical treatment is recommended only in the case that is impossible to avoid it.

Chapter 3　Basic Principles of Powder Metallurgy of Nanomaterials

3.1　General Information

Nanomaterials are materials containing structures whose geometric dimensions do not exceed 100 nm in at least one direction, and possess qualitatively new properties, functional and operational characteristics.

Nanomaterials can provisionally be divided into four classes:

① Materials in the form of solids, whose size in one, two or three dimensions does not exceed 100 nm (nanopowders, nanofibers, nanotubes, thin films, nanowires, etc.). These materials may contain from one structural element or crystallite (for powder) to several layers (for films). This class is classified as nanomaterials with a small number of structural components or nanomaterials in the form of nanoparticles.

② Materials in the form of small-sized articles with a characterizing size in the range of 1 μm—1 mm (wire, tape, foil). Such materials already contain a significant number of structural elements, and can be classified as nanomaterials with a large number of structural components (crystallites) or nanomaterials in the form of microproducts.

③ Massive (or volume) nanomaterials with dimensions of products in the macro region of more than 1 mm. Such materials consist of a very large number of nanoscale components and are in fact polycrystalline materials containing structure components with dimensions of 1—100 nm.

④ Composite materials containing components of nanomaterials. Components may include nanomaterials classified in class ① (nanoparticles and/or nanofibers, products with modified ion implantation by surface layer or thin film) and class ② (composites reinforced with fibers and/or particles with nanostructure, materials with modified nanostructured surface layer or coating).

Nanotechnologies—the processes of obtaining nano- or ultra-disperse materials, as well as technological processes of obtaining instruments, devices and structures using nanomaterials. Nanotechnologies are divided into evolutionary technology and revolutionary technology. The first ones are based on the use of already existing nanoproducts (nanopowders, carbon nanotubes, fullerenes, etc.), while the second ones should be created as a result of fundamental research processes of self-assembly and self-organization, following the "bottom-up" direction[70].

"*Bottom-up*" nanotechnology is a technology for obtaining nanostructured materials, in which the formation of nanoparticles from atoms and molecules is realized, that is, the aggregation of the original elements of the structure up to particles of nanometer size is achieved. Techniques of this type include methods used to produce isolated nanoparticles, nanopowders and compact nanomaterials—gas phase synthesis with following condensation of vapor; plasmochemical synthesis; deposition from colloidal solutions; chemical and physical deposition of films and coatings from the gas phase (CVD and PVD processes), plasma or liquid solutions on the substrate; electrodeposition of films and coatings; thermal decomposition (pyrolysis); detonation synthesis[71-72].

"*Top-down*" nanotechnology is a technology for producing nanostructured materials in which nanometer size of particles is achieved by grinding larger particles, powders or grains of a solid. The "Top-down" nanotechnology includes the following methods to obtain compact nanomaterials and nanopowders from bulk blanks: crystallization of amorphous alloys, severe plastic deformation, electric explosion[71].

Nanotechnologies are qualitatively different from traditional disciplines, because on such scales range, macroscopic technologies for dealing with matter are often inapplicable, and microscopic phenomena, negligibly weak on large scales, become significant (properties and interactions of individual atoms and molecules or aggregates of molecules, quantum effects)[73].

Nanostructure is a set of nanosized objects of artificial or natural units, whose properties are determined not only by the size of structural components, but also by their mutual arrangement in space.

The classification of nanostructures is based on their dimension: 0D, 1D, 2D, 3D, which corresponds to zero-, one-, two-, and three-dimensional structures. All nanostructures can be built from constructional components (blocks), which can be divided into three classes 0D, 1D and 2D. 3D blocks can also act as constructional blocks for structures of a higher hierarchical level. Nanostructures are marked as NC (kD$lmn\cdots$), where k is the dimension of the nanostructure as a whole; l, m and n are the dimensions of the construction components. By definition, $k \geqslant l$, m, n, with k, l, m, n only being integer numbers from 0 to 3. These conditions lead that there are only 3 elemental varieties of nanostructures (0D, 1D, 2D), 9 single classes of kDl type, built from construction components of one grade, 19 binary classes of type $kDlm$, built from construction components of two grades, as well as many triples, quad, and so on classes[74].

Both already known and not yet synthesized structures fit into this classification. Thus, granular (or particulate) systems consisting of isodiametric particles or granules are referred to as 3D0, 2D0, 1D0, 0D0. Fiber (or fibrillar) systems consisting of fibers with arbitrary length with an average cross-section diameter of no more than 100 nm are referred as 3D1, 2D1, 1D1. One can present a nanostructured 3D1 system, which consists of one continuous fiber, which is, for example, coiled up. Film (or laminar) systems consisting of thin films, plates, flakes, etc. with thickness of no more than 100 nm, are referred to as 3D2, 2D2. The films have size in two dimensions, comparable to the size of the body as a whole. Also, the nanostructured laminar system consists of a single film rolled up into a multi-layer 3D2 roll. This classification of nanostructures allows its expansion to additional geometric and real features[74].

The main method of obtaining *consolidated nanomaterials* (compacts, films, coatings) is the methods of powder metallurgy, severe plastic deformation, controlled crystallization from the amorphous state, and various methods for films and coatings. Nanograins (nanocrystals) of such materials are not in an isolated (that is, in the form of separate formations) or weakly bound (for example, nanoparticles with protective polymeric shells) state, but in a consolidated state. The strength of intergranular interlayers in consolidated nanomaterials is rather high.

Powder nanotechnologies form a significant market share of all nanotechnologies, especially at the initial stage of innovation development in this field. In the technological chain "synthesis of nanopowders" the final product (nanomaterial) turns out to be very sensitive to all stages of its prehistory, starting with the quality parameter of the nanopowder, including strict adherence to all parameters of subsequent processing. High target properties are achieved by the formation of certain structures of nanomaterials from nanopowders. The development and implementation of such methods are actually the subject of technological research in powder nanotechnology.

Optimization of the structure of nanomaterials in order to give them high strength values with satisfying plasticity is the quintessence of many searches in the field of science of nanostructured materials. The main methods of increasing the plasticity of nanomaterials based on metals and alloys are:

① creation of bimodal structures in which nanocrystalline matrix provides high strength, and availability of larger inclusions facilitates to achieve acceptable plasticity;

② obtaining multi-phase compositions;

③ formation of twin structures;

④ production of dispersion-strengthened alloys;

⑤ use of "TRansformation Induced Plasticity" (TRIP) and "TWinning Induced Plasticity" (TWIP) effects;

⑥ dynamic annealing at low temperatures;

⑦ increasing the coefficient of speed sensitivity of the flow stress up to the values inherent in superplastic alloys, by repeated processing including the methods of intense plastic deformation;

⑧ use of high-speed and ultra-high-speed methods of heating in the consolidation of nanoparticles in order to prevent their growth;

⑨ use of methods of consolidation of nanomaterials, ensuring the complete absence of pores and discontinuities due to the occurrence of stress concentration which is the positive effect of nanostructure.

A new direction in modern science of materials, whose principles are aimed at improving the properties of nanomaterials, is *grain-boundary engineering*. Creation of mostly high-angle boundaries with a non-equilibrium structure or formation of grain-boundary segregations and precipitations provides nanomaterials with unique combination of high strength and plasticity and, as a result, obtaining materials with high fatigue resistance, durability and improved toughness—properties that are necessary for products with long service life.

3.2 Methods of Obtaining Nanopowders

Nanopowder is a) a solid powdery substance of artificial origin, containing nano-objects, aggregates or agglomerates of nano-objects, or their mixture (according to ISO); b) nanoparticle ensemble; c) powder, whose size of all particles is less than 100 nm.

There are several parameters of nanopowders that are important for assessing their quality:

① Average size of particles.

② Size distribution of particles (granulometric composition), which significantly depends on the method of their production. If the average nanoparticle size is 30 nm, the powder particles up to 200 nm and even 1 μm will be present while using the electric wire explosion method. The method of laser evaporation and condensation gives a narrow distribution (a narrow spectrum of particle size distribution).

③ The degree of agglomeration of particles (weak agglomeration—the bond of particles due to interactions of the Van der Waals type, strong aggregation is characterized by strong interparticle bonds). The most valuable nanopowders for creating new materials are weakly aggregated powders.

Nanopowders are also characterized by the chemical composition of the particle, the composition over the cross section for core-shell particles, the morphology of particles, chemical composition of the surface, crystalline structure of the nanoparticles, moisture content and other adsorbates, friability (flowability), bulk density, color[75].

Methods of obtaining nanopowders may be divided into two groups:

1. Technologies based on chemical processes

① Chemical vapor deposition (transfer through the gas phase followed by deposition);

② High-energy synthesis (detonation, plasma-chemical);

③ Deposition from solutions (chemical deposition, sol-gel method, liquid phase reduction, hydrothermal synthesis, microemulsion method, cryochemical method);

④ Decomposition of unstable compounds (thermal, radiation);

⑤ Reduction processes (hydrogen reduction of metal compounds, chemical and metallurgical method).

2. Technologies based on physical processes

① Physical vapor deposition (thermal induction evaporation, electric arc, electron beam, laser heating; explosive evaporation—electric conductor explosion);

② Spraying of the melt (using a water-borne disk or drum, impact spraying, electrodynamic spraying);

③ Mechanical grinding (grinding in mills, countercurrent grinding in a fluidized bed).

The division into chemical and physical methods is very conditional. Thus, chemical reactions play a large part, for example, during evaporation in the medium of the reaction gases. At the same time, many chemical methods are based on physical phenomena (low-temperature plasma, laser or electron radiation). Chemical methods are usually more universally applicable and productive whereas it is easier to control the size, composition and shape of particles by physical methods.

3.3 Size Effects in Nanoparticles

Size effects in nanoparticles appear in a complex of phenomena associated with changes in the properties of a substance due to:

① direct change of particle size;

② increase in contribution of interfaces to the properties of the system with decreasing particle size;

③ commensurability of particle size with physical parameters, having the dimension of length, the influence of small grain size and, accordingly, the developed surface of the nanoparticles on the properties of nanomaterials.

With a decrease in the grain size, the volume fraction of the boundaries ΔV_{if} increases, which is determined by the equation

$$\Delta V_{if} = \left(\frac{D-t}{D}\right)^3 \tag{3.1}$$

where t—boundary thickness (~ 1 nm);

D—characteristic size, e.g., grain size, nm.

Volume fraction of grain boundaries ΔV_{bg} is determined by the equation

$$\Delta V_{bg} = \frac{3t(D-t)^2}{D^3} \tag{3.2}$$

Volume fraction of triple joints ΔV_{tj} is determined by the equation

$$\Delta V_{tj} = \Delta V_{if} - \Delta V_{bg} \tag{3.3}$$

With a decrease in the grain size from 1 μm to 2 nm, the volume fraction of the intergranular component increases from 0.3% to 87.5%. The volume fractions of intergranular and intragranular components are equal (that is, 50% each) with a grain size of about 5 nm. The volume fraction of triple joints increases significantly when the grain size is less than 10 nm.

In nanoparticles, there may occur phases that do not exist when the substance is in a massive state. With the decrease in particle size, there increases the contribution of the surface energy F_s in full particle free energy $F = F_v + F_s$, where F_v is the fraction of free energy due to volume contribution. If phase 1 is stable in bulk samples at a certain temperature, that is $F_{v(1)} < F_{v(2)}$, then with decreasing size, taking F_s into account, the inequality $F_{v(2)} + F_{s(2)} \leqslant F_{v(1)} + F_{s(1)}$ may be true, and with sufficiently small particle sizes the phase 2 will be stable[76].

Since the surface energy is a noticeable value compared to the bulk energy, it follows from the above inequality that to reduce the total energy of the system, the deformation of the crystal can be utilized to decrease the surface energy. Such a reduction can be realized by changing the crystal structure of the nanoparticle compared to the crystal structure of the bulk sample. The surface energy is minimal for close-packed structures; therefore, FCC and HCP structures are the most preferable for nanocrystalline particles[76].

The transition from massive crystals to nanoparticles is accompanied by a decrease in the interatomic distances and periods of the crystal lattice. This is due to the uncompensated interatomic bonds of the atoms of the surface of the particles compared to the atoms located inside it. As a result, the distances between the atomic planes near the particle surface, that is, the surface relaxation, decrease. Surface relaxation captures several surface layers and causes a change in the volume of the particles. For nanoparticles, the surface relaxation is maximal on the surface, decreases from the surface to the center of the particle, and under certain conditions it can be oscillating. Depending on the lattice periods and crystal size, surface relaxation can not only reduce, but also increase its volume.

The change in thermodynamic characteristics of nanoparticles (heat capacity, thermal expansion, melting point) in comparison with a bulk substance is determined by the change in the form of the boundaries of the phonon spectrum due to the presence of a large number of atoms on the surface of nanoparticles. Additional modes appear in the phonon spectrum of nanomaterials and an increase in the amplitude of atomic vibrations is observed. From the side of low-frequency oscillations, the spectrum is limited to the minimum frequency ω_{min}:

$$\omega_{min} = \frac{c}{2D} \qquad (3.4)$$

where c—sound speed.

There is no such limitation in bulk materials[76].

Magnitude ω_{min} depends on the properties of the matter, shape and size of the particle. It is expected that with decreasing particle size, the phonon spectrum will shift to high frequencies. The characteristics of the vibrational spectrum of nanoparticles are reflected in the heat capacity. For example, the heat capacity of a Pd nanoparticle is higher than the heat capacity of bulk Pd[76].

A noticeable decrease in the melting temperature is observed when the nanoparticle size is less than 10 nm. A decrease in the melting point of nanoparticles Sn, Pb, In, Ag, Cu, Al, Ga, Au was experimentally observed. One of the most interesting properties of nanoparticles is the presence of a complete magnetic moment of a cluster consisting of nonmagnetic atoms. For example, the Re cluster demonstrates a distinct increase in the magnetic moment if it has less than 20 atoms[76].

The size effects of optical properties are significant for nanoparticles whose size is noticeably smaller than the wavelength and does not exceed 10—15 nm. The differences in the absorption spectra of nanoparticles and bulk materials are due to the difference in their dielectric constant. For nanoparticles with a discrete energy spectrum, it depends both on their size and on the radiation frequency, with the latter dependence being not monotonous, but oscillates due to transitions between electronic states. A decrease in the size of semiconductor nanoparticles is accompanied by a shift of the absorption band toward the high-frequency region, which is manifested in the blue shift of the exciton absorption band. With a decrease in the size of the nanoparticles, a shift of the smallest absorption energy, called the absorption boundary, towards higher energies, and an increase in the absorption intensity occur.

The ability of nanoparticles to react with other substances depends on their size, which is caused by the dependence of the electronic structure of the nanoparticle on its size. Thus, Al_{12}, Al_{24}, Al_{19} and Al_{20} interact intensively with oxygen while the Al_{13} and Al_{23} clusters practically do not interact therewith, which indicates the dependence of the reactivity of the clusters on the number of atoms therein. This fact is of great importance for designing the catalysts. The high catalytic activity of gold nanoparticles with a size of less than 3—5 nm, having an icosahedral structure, has been realized[77].

With a decrease in the size of a nanoparticle, dislocations are removed in an increasing part of its volume. Experimental studies of the structure of small particles showed the absence of dislocations therein. This is indirectly confirmed by a significant increase in the pressing stress with a decrease in the particle size of the powders. However, dislocations in polycrystals with coherent grain boundaries are stable. Thus, dislocations in the bulk of the grains were found in compacts from Pd powder with a particle size of 20 nm and a Mo powder with a particle size of 10 nm[76].

3.4 Consolidation of Bulk Nanomaterials

To achieve a nanostructured state in a bulk material by powder metallurgy, it is used that both nanopowders with particle sizes of less than 100 nm and powders of a much larger size with a nano-microcrystalline structure obtained by mechanical alloying. Powders of amorphous alloys, which are subjected to controlled crystallization during sintering, can also be used.

3.4.1　Pressing of Nanopowders

Obtaining compacts from nanopowders with uniform density throughout the volume is challenging due to poor compressibility. The physical reason of poor compressibility of nanopowders is interparticle adhesive forces, whose relative magnitude increases sharply with decreasing particle size, and therefore the interparticle friction in the pressed powder body significantly increases.

It is necessary to use pressure sufficient to redistribute the particles in the volume to achieve a sufficiently high density. Since a high number of nanoparticles is present in a unit volume, the sliding friction resistance increases due to a large amount of interparticle contacts. In addition, nanoceramic powders contain solid agglomerates (strongly bound particles) that must be destroyed in order to redistribute the particles during their pressing.

Due to the high adsorption capacity of nanopowders, affecting their compressibility, the amount of absorbed gases in them can reach 20%. This leads to special requirements for storage, technological preparation and handling of nanopowders at all stages of powder technology[78].

Another problem is to preserve the nanostructures of nanocompacts prior to sintering, that is, to prevent the intensive agglomeration of nanoparticles at high compaction pressures, in order to preserve the nucleation centers of nanograins and to avoid grain growth during sintering. For dust-like nanopowders, an even more difficult task is to achieve uniform density of compacts with a given shape and prevent the formation of macro defects from internal stresses therein because of high elastic effect, which is especially typical for rigid ceramic nanopowders[78].

It should be noted that the dispersion of powders has a much higher effect on their compressibility than physical and mechanical properties. Thus, nanopowders of plastic nickel and brittle silicon nitride are pressed almost identically, despite the fundamental differences in their properties. Nevertheless, the type of nanopowders (plastic or brittle) is of great importance for the process of their pressing and the development of optimal methods for their compaction.

When selecting methods and schemes for pressing nanopowders, the following factors should be considered:

① Deformations in compressed powder materials develop in a substantially non-uniform manner. Plastic deformations are localized in shear bands, in which the degree of strain accumulation and, accordingly, the degree of compaction and hardening of the powder material are very high. The deformation in shear bands is much higher than that outside the shear bands.

Chapter 3 Basic Principles of Powder Metallurgy of Nanomaterials

② Low compressibility and high compressive resistance of compacts from nanopowders are caused by the fact that, due to localization of deformation in shear bands in a pressed material, a structure is formed consisting of a high-density, high-strength and high-modulus "frame" that absorbs almost the entire load, and unconsolidated highly porous material is enclosed inside cells of such a frame.

③ The high interparticle adhesion of nanopowders and the low bulk density associated therewith leads to the fact that the effect of hardening of the material due to the formation of a high-strength "composite structure" in the process of its pressing is much more apparent when compacting such powders than when compacting powders of medium dispersion.

④ Since the formation of shear bands, from which a high-strength "frame" is formed in pressed nanomaterials, occurs along non-orthogonal sliding surfaces, the structure of such compacts is anisotropic in strength and is determined by the direction and parameters of the applied load.

Due to the peculiarities of nanopowders (agglomeration, high sorption activity, high interparticle and near-wall friction during pressing, lower bulk density), traditional pressing methods do not provide uniform density of compacts, lead to local density gradients therein, high internal stresses, large elastic after effect and, as a result, cracking or destruction of the compacts. In view of the aforesaid, it was necessary to develop special methods for compacting nanopowders. In most cases, they were methods modified from known methods of compacting powders, which often meant optimization of physicochemical regimes for compaction of nanopowders. However, there were developed compaction methods that use the specific properties of nanopowders (superplasticity in a certain temperature range, metastability of the structural phase state, correlation of physical properties with the size of nanoparticles)[78].

Cold static pressing in closed molds is the simplest method from the point of view of technological implementation and universal for pressing products of various shapes from powders of any composition by means of compaction. The disadvantage of this method is the uneven distribution of properties by volume of the compacted product. As already noted, the cause of the density drop across the height of the compact is the loss of pressing force to overcome the forces of near-wall friction, which are quite significant especially for nanopowders with a high specific surface.

Isostatic pressing is the most preferred method for forming samples from nanopowders. At isostatic pressing (modifications—hydrostatic, gas-static, quasi-hydrostatic), the conditions close to all-round compression are observed, which is extremely important for nanopowders with poor compressibility. At isostatic pressing, the powder is placed in an elastic form, which is influenced by external pressure, transmitted by gaseous or liquid medium. The resulting compacts have almost uniform density (sometimes, depending on the dimensions of the products, it is somewhat less in the internal volumes of the compact) and do not have anisotropy of properties. The disadvantage of the method is the complexity and high cost of equipment as well as difficulties in obtaining the exact form of compacts.

Hydrostatic pressing is a form of isostatic pressing, usually performed at room temperature. The essence of the method is that the powders (including compacts of complex geometric shape) are poured into a shell of elastic material with a thickness of 0.1—2.0 mm, evacuated and placed into a working chamber of a hydrostat, where the liquid is then injected (oil, glycerin, more often water) under pressure of 100—1 200 MPa.

Quasi-hydrostatic pressing is a simplified version of hydrostatic pressing. The powder is placed in an elastic form, whose pressing is carried out with unilateral or two-sided application of pressure on conventional pressing equipment. The cover material (rubber, epoxy resins, polyurethane, etc.) should behave like a liquid under pressure, transferring pressure in all directions and have certain elasticity, not adhering with powder.

Hot pressing method provides technology to obtain ceramic products with a small density difference in volume. This method, which has the effect of superplasticity at elevated temperatures, is effective for production of certain types of nanoceramics. This method can significantly reduce the cost of products. Eliminating the machining operation is the way to reduce the cost. However, the application of this method is limited by the high requirements for the material of the molds, which must be heat-resistant, and resistant to creep and economical. The advantages of the hot pressing method are the good strength properties of the products, minimum or absent tolerances in the size of the workpieces, and reduction of the technological cycle due to the combination of sintering and pressing operations. Disadvantages of the method is low resistance of the molds and low productivity of the process.

At *dynamic pressing methods*, compaction is carried out by shock waves. Electrohydraulic discharge of a battery of high-voltage capacitors, the energy of a pulsed magnetic field, compressed and combustible gases, as well as liquids and explosives are used as energy carriers. These methods include isothermal stamping, magnetic-impulse, explosive, hydrodynamic, shock pressing, electro-consolidation, and vibratory molding. These compaction methods are carried out at high dynamic (pulse) pressures (more than 1 GPa), which determine the special requirements to the strength of the dies and molds. As a rule, there are required special operations for preliminary preparation of powders for destruction of agglomerates, removal of absorbed gases, evacuation and annealing at elevated temperatures. Due to the pulsed nature of the impact, unrelaxed stresses are created in compacts, so annealing is often necessary before removing compacts from molds.

The basis of the *magnetic pulse compaction* of nanopowders is the electromechanical conversion of the energy of the primary capacitive storage device into the kinetic energy of the press tool, which compresses the powder. At magnetic pulsed nanopowder compression high impulse pressure determines the compaction of nanoparticles. The effect of large mechanical impulse of particles results in a significant decrease in the role of interparticle interaction, which can effectively increase the particle mobility, and reduce the internal friction at a macro level. This makes it possible to obtain compacts from nanopowders with a higher density, and the role of this effect is enhanced with a decrease in the average particle size in the powder. Due to the adiabaticity of the process, soft pulsed pressing is characterized by a significant pulsed heating of the pressed powder, which improves its compressibility and is able to stimulate structural phase transformations. Due to the rapidity of pulsed pressing, it is possible to preserve metastable structural and phase states, which are preferable for forming the properties of a bulk nanomaterial. Magnetic and pulse method of pressing is used to obtain parts of various shapes, as a rule, without further machining. The porosity of the products obtained by this method was found to be inhibited[79].

Ultrasonic quasi-resonant pressing is a method of compaction (without plasticizer) of difficult-to-form nanopowders with ensuring minimal density drops in the volume of products. During dry pressing of ceramic nanopowders under quasi-resonant conditions with the influence of ultrasound, an oscillatory displacement of particles or agglomerates of nanopowders occurs. The amplitude of sound vibrations supplied to the mold is close to the average particle size. Thus, the size factor of nanopowders is used, when, even at a sufficiently large pressing depth, the vibrational displacements of the nanoparticles of the powder from the action of damped ultrasound are commensurable with their small size. Optimal ultrasonic compaction ideally implies monodispersity of the powder. In reality, the powder is always polydisperse; therefore, at almost any amplitude of ultrasonic vibrations, there will be such a range of particle sizes that will correspond to their vibrational displacement at a certain depth of pressing from the vibrating surface of the mold. A positive point in using the ultrasound is the reduction of the static pressure required for compacting the product, with an increase in the amplitude of oscillations of the tooling. With a decrease in particle size, it is necessary to increase the frequency of vibrations, and for powders with non-plastic particles it is also necessary to use higher frequencies, but with a smaller amplitude of vibrations.

3.4.2 Sintering

The formation of nanostructures in the material produced by the methods of powder technology "bottom-up" is possible only in case of using nanostructured powders as nucleation centers of nanograins. Necessary conditions for sintering compacts of nanopowders, providing the suppression of grain growth (recrystallization) and obtaining high-density nanoceramics are:

① high compacting density (not less than $0.7\rho_{th}$), contributing to the rapid sintering processes.

② relatively low sintering temperature (no more than $0.5T_{mel}$).

③ high compaction speed that can be achieved only with small size of pores at all sintering stages, i.e. pore and grain sizes should be controlled during sintering. In this case, the speed of compaction is determined by the current (instantaneous) pore size, but not by their initial size. Small pore size throughout sintering is also a critical factor for forming the final grain size.

④ uniform distribution of small sized pores.

⑤ narrow size distribution of pores.

⑥ preservation of the nanostructured state in the compressed powder before sintering, preventing plastic deformation of nanopowder particles during pressing.

⑦ the use of high-energy methods of consolidation associated with the use of static and dynamic high pressures, electropulse loads, etc.

⑧ introduction of insoluble additives to the initial powder (mixture), localized at the grain boundaries and preventing their aggregation.

The problem of limiting high temperatures during sintering of compacts from nanopowders is solved through the use of various activation methods that provide low porosity at lower sintering temperatures, namely:

① high-speed microwave heating (for example, with an increase in the heating rate from 10 ℃/min to 300 ℃/min, the sintering temperature of TiO_2 nanopowder decreases from 1 050 ℃ to 975 ℃);

② step controlled sintering;

③ plasma activated sintering;
④ carrying out sintering in vacuum or reduced environments (for metal powders).

Spark Plasma Sintering (SPS) is an ultra-modern sintering method in spark plasma, which can be used for sintering all ceramic and metal powders (including nanopowders) virtually in the shortest possible time (ensuring high density), making it possible to produce substantially improved materials, up to completely new materials[80-81].

SPS technology is a consolidation method based on traditional hot pressing (see Fig. 3.1). SPS installations consist of a water-cooled chamber, a hydraulic pressing system, and a computer control system, which controls the temperature, force, and vacuum (or gaseous medium) inside the chamber. The main difference from the traditional hot pressing is that with the SPS technology there is no heating components and traditional thermal insulation of the chamber. Instead, through a special energy source, high voltage current is supplied to the dies of the water-cooled machine, so that they simultaneously act as electrodes and directly conduct high-voltage current through the mold and the pressed powders in it. This design allows to perform a uniform volumetric heating (Joule) of the mold and the powders therein. Because of this, even at high heating rates, relatively small gradient temperature arises, while at traditional sintering methods, because of temperature gradients, there is a risk of damage, and therefore only average heating rates are set and, as a result, longer exposure times for subsequent homogenization is used. With SPS technology, the heating power is not only evenly distributed across the volume of the powders at macroscopic level, but also supplied exactly to the regions where energy is required for sintering the regions of poor electrical contact, that is, contact points of the powder particles at microscopic level. The advantages of SPS technology include:

① uniform distribution of heat throughout the volume of the sintered product;
② high density and controlled porosity;
③ minimum grain growth;
④ minimal effect on the microstructure;
⑤ no pre-treatment by pressure and bonding;
⑥ uniform sintering of homogeneous and inhomogeneous materials;
⑦ evaporation of existing impurities;
⑧ manufacturing of a part directly in the final form;
⑨ short cycle time;
⑩ low production costs;

⑪ convenience of use.

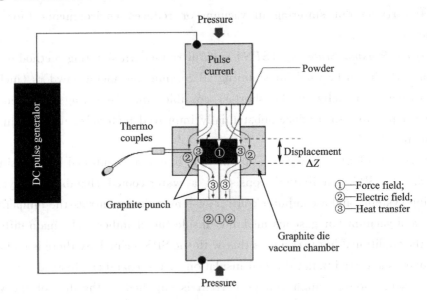

FIGURE 3.1　Schematic of the SPS process[81]

Sintering using *microwave radiation* of ceramic nanomaterials is based on high-frequency heating of the sintered product with millimeter-wave radiation (frequency range is usually 24—84 GHz). Volumetric absorption of microwave energy provides simultaneous uniform heating of the product since the heating rate is not limited by thermal conductivity, as in traditional methods. This makes it possible to obtain sintered ceramics with a uniform microstructure. For example, microwave sintering of compacts from Al_2O_3, pressed from a nanopowder with an average particle size of 26 nm and having a relative density of 52% made it possible to obtain a finished sintering product with a density of 99% and an average crystallite size of 80 nm. The sintering temperature was 1 500 ℃[82].

Hot gas extrusion, which includes compact production, is a technology for obtaining bulk blanks (long metal products with a nanocrystalline structure), sintering it in a reducing environment to remove the oxide film from the surface of metal particles and hot pressing. In this case, sintering in a reducing environment is carried out until the closed porosity in the preform is reached, and hot pressing is carried out by extrusion with an inert gaseous medium with local heating of the deformation zone below T_{rec}.

Thus, Ni nanopowders with an average particle size of 72 nm were hydrostatically compressed in elastic shells and compacts with a density of 60% were obtained, which were then sintered in hydrogen in a two-step mode at temperatures of 500 ℃ and 700 ℃ (exposure at the indicated temperatures for 2 h). Sintered samples with a density of 93.5% were subjected to extrusion through a matrix with an outlet of 2 mm. The degree of deformation was 92.5%. The relative density of nickel samples as a result of processing by gas extrusion in the temperature range 800—900 ℃ increased from 93.5% to 98.5%. The microstructure of the nickel sample was characterized by sufficient uniformity, small grain size and the absence of pores[83].

Chapter 4　Structural Materials

As mentioned earlier (see Chapter 1), in aerospace engineering, despite the variety of grades of metallic materials, they are literally classified into several main groups. Given that powder metallurgy is rather a young branch of science and technology, powder materials are being introduced, including aerospace technology, on a competitive basis. This means that in order to apply the powder technology, the manufactured products (parts) should be better at least by one performance indicator than those manufactured by the traditional method. Or, in a pinch, not to be worse, but the products (parts) manufactured by powder technology have better economic indicators, that is, to be cheaper in manufacturing. This is the main logic for using the structural powder parts in aircraft.

4.1　Powder Metallurgy Aluminum Alloys

As was shown earlier, *aluminum alloys* continue to confidently take the leading place in the aircraft industry[84]. At present, the research conducted in the field of aluminum alloys is directed for searching for both new alloy systems and new technological methods for production of blanks and finished parts. In particular, much attention is paid to producing aluminum-based composites, which will be discussed later.

Aluminum is a chemical element of the third group of the periodic system of Mendeleev: a) the chemical symbol is Al; b) in solid state it does not undergo polymorphic transformations; c) crystal structure is face-centered cubic with a crystal lattice parameter of 0.404 9 nm; d) atomic number is 13, and atomic radius is 143 pm; e) atomic mass is 26.981 5; f) melting temperature $T_{mel} = 660\ °C$; g) evaporating temperature $T_{ev} = 2\ 500\ °C$; h) density $\rho = 2\ 710\ kg/m^3$; i) elastic modulus $E = 71\ GPa$; j) tensile strength $R_m \leqslant 60\ MPa$; k) electrical resistivity $\rho = 2.62 \times 10^{-8}\ \Omega \cdot m$; l) linear coefficient of thermal expansion $\alpha = 23.9 \times 10^{-6}\ K^{-1}$ in the temperature range of 20—100 °C; m) Poisson ratio is 0.34; n) thermal conductivity $\lambda = 226\ W/(m \cdot K)$[85].

The wide use of powder metallurgy aluminum alloys is determined by the low cost of raw materials, high level of physical and mechanical properties and corrosion resistance, low density, good workability, including cutting property.

The low energy intensity of production of aluminum alloys is determined by the low values of cold-pressing pressure, not high sintering temperatures, which do not require powerful technological equipment[86].

The first aluminum materials that were manufactured using powder metallurgy technology were the so-called SAPs—that is, materials sintered from aluminum powder (Sintered Aluminum Powder (SAP)).

The main advantage of these materials was the preservation of relatively high tensile strength in the temperature range of 300—500 ℃, while the cast high-temperature aluminum alloys are operated at temperatures of up to 200 ℃. The development of the first industrial SAPs refers to the 50s of the twentieth century.

The physical and chemical features of aluminum and aluminum oxide (Al_2O_3) are the basis of the heat-resistant properties of SAP. Aluminum is known as a soft, fragile, lightweight, low-melting silver-white metal, which oxidizes easily and quickly in air at room temperature with the formation of a thin dense oxide film on the surface, which prevents further oxidation. Aluminum oxide is one of the hardest substances in nature (second only to diamond), with $T_{mel} = 2\,500$ ℃. Moreover, Al_2O_3 almost does not dissolve in aluminum. Thus, aluminum oxide can be an effective hardening phase for aluminum alloys, in terms of strength properties and ease of addition.

The powder is made by spraying liquid aluminum into water, producing oxidized aluminum particles in the form of plates (flakes) with dimensions $d = 10$—45 μm, $h = 1$ μm. By changing the temperature of the sprayed melt, one can adjust the content of Al_2O_3 in the powder from 6% to 15%. Therefore, SAP alloys with different content of the hardening phase, and, hence, the properties can be obtained. SAP is used in lightly loaded structures operating up to a temperature of 500 ℃. The use of SAP in highly loaded structures is limited by their low fatigue properties—aluminum oxide particles provoke the formation of fatigue cracks.

From the atomized powder, we can obtain materials SAP 1 (6% Al_2O_3), SAP 2 (9% Al_2O_3) and SAP 3 (15% Al_2O_3), respectively. The mechanical properties are presented in Table 4.1. It is shown that by hardening aluminum with its oxide, it is possible to increase the tensile strength by 5—7 times. Given the significant increase in strength, many researchers tried to make semi-finished products from traditional aluminum alloys using SAP technology. Experiments conducted on alloys of systems 2xxx (Al-Cu-Mg) and 7xxx (Al-Zn-Mg Cu) showed that there is no increase in strength in alloys compared to pure aluminum. Despite the negative results of the experiments, it was shown that by using powder metallurgy technology it is possible to obtain a fine-grained structure in the aluminum alloy, which is prone to collectively recrystallization, which is unattainable in the foundry method of alloy production.

TABLE 4.1 Mechanical and physical properties of samples made by SAP

Samples	R_m/MPa	$R_{p0.2}$/MPa	A/%	HB/(kg·mm^{-2})	E/GPa
SAP 1	330	230	6	95—100	75
SAP 2	440	260	3	120	77
SAP 3	450	330	3	120	78

It is well known that the smaller the grain size, the higher the mechanical properties of the final alloy. For strength properties, there is the Hall-Petch law describing the effect of the grain size:

$$R_m = R_m^0 + \kappa d^{-1/2} \qquad (4.1)$$

where R_m and R_m^0 —tensile strength of poly- and mono-crystals;

κ —stress transfer coefficient at grain boundaries;

d —grain size.

In parallel with the increase in strength, a decrease in grain size increases the ductility of the alloy. In addition, the smaller the size of dispersed particles (hardening), the less they serve as sources of crack initiation. Furthermore, the reduction in grain size and size of dispersed second phase in aluminum alloys often improves corrosion resistance. These prerequisites led to appearance of powder metallurgy aluminum alloys in aircraft practice.

Obtaining aluminum alloy powders by spraying (granule metallurgy—the term introduced by the member of the academy A. F. Belov[87]), significant high cooling rates (10^4—10^6 ℃/s) can be achieved, which makes it possible to achieve a number of advantages with respect to traditional technologies, namely:

① Depending on the cooling rate, it is possible to obtain a regulated grain size up to several microns by atominzation, since during crystallization the growth time of the nucleus is reduced to minimum, and the particle is in micro-size.

② In the process of rapid crystallization, diffusion processes do not have time to occur and the phase rule does not work, which makes it possible to obtain supersaturated solid solutions and avoid harmful segregations which occur in cast alloys during their re-alloying.

③ Some elements that are not soluble in the solid state, but soluble in the liquid (for example, Co), uniformly disperse in the powder particles during the rapid solidification process.

④ During high-speed cooling, unstable nonequilibrium phases and atomic "clusters" that are not formed during slow cooling, which are effective hardening phases, are formed.

⑤ Rapid cooling can suppress the release of coarse equilibrium phases, deteriorating the mechanical and corrosion properties.

Based on the above, we can distinguish two main areas in producing powder metallurgy aluminum alloys:

① Production of alloys based on the known alloying systems, obtaining advantages due to modifying structure and changing of phase transformations kinetics.

② Production of new alloying systems, taking into account the possibility of introducing alloying elements beyond the solubility limit and elements insoluble in the solid state.

The first direction is simpler for introduction of alloying elements into products, since the advantages of the technology comparing with the traditional are immediately visible and there is no need to overcome the "caution" and "conservatism" of aerospace engineering that exist in practice. So, almost all high-strength aluminum alloys can be obtained by two technologies: ingot casting with subsequent deformation redistribution and powder metallurgy method according to the below technological schemes (see Fig. 4.1).

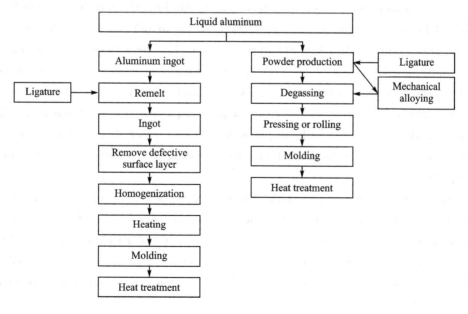

FIGURE 4.1　Technological production scheme of aluminum alloys

Production of powder metallurgy aluminum alloy is carried out in two ways—

by two-fluid atomization (gas or liquid) and centrifugal atomization under the action of centrifugal forces of a rotating crystallizer into a gaseous medium or water[88]. It was previously established that if you do not protect the surface of aluminum particles, the oxides formed, even in a small amount, adversely influence the mechanical properties of the alloy and block its hardening by heat treatment. On the other hand, when the alloy is atomized by air, easily oxidizable alloying elements (for example, Li) burn out. In this regard, the atomization of the alloys is carried out with inert gases. Fig. 4.2 gives examples of aluminum powders which were gas atomized under nitrogen[89].

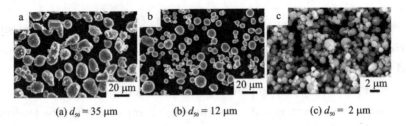

(a) $d_{50} = 35$ μm (b) $d_{50} = 12$ μm (c) $d_{50} = 2$ μm

FIGURE 4.2 SEM micrographs of Al powder

During centrifugal atomization into the water (more productive process), moisture is absorbed on the powder particles, which cannot be removed by simple drying. Therefore, to obtain high-quality semi-finished products from powder metallurgy aluminum alloys, it is necessary to produce the powder immediately, without bringing it into the air, to place it in an airtight container and degas it (degassing removes absorbed moisture and gases). The process of degassing of aluminum powder is the most energy-intensive and labor-intensive process in the production of semi-finished products and it dramatically increases the cost of powder alloys. There are several technologies for powder degassing but given that when aluminum is heated with the presence of bound water, hydrogen is produced, which can explode, the method of vacuum degassing in a welded container is most acceptable (see Fig. 4.3).

Degassing is carried out at temperature of 300—400 ℃ to remove the adsorbed moisture. It was established that if the degassed powder is treated with argon, it loses the tendency to re-sorb moisture. Therefore, in order to obtain high-quality, non-porous semi-finished products, repeated evacuation is carried out followed by filling in the container with dry argon. After degassing, the container is transferred to pressing.

The most effective pressing method, which allows to obtain 100% density of compacts, is Hot Isostatic Pressing (HIP). The obtained compacts, after removing

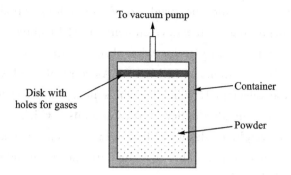

FIGURE 4.3 Technological scheme for vacuum degassing of aluminum powders

the container, can be converted into any semi-finished products by traditional methods of hot deformation.

Taking into account that using powder technology it is possible to increase the level of mechanical properties of serial deformable aluminum alloys, the interest of researchers is naturally focused on those, where the advantages are most obvious. Modern deformable high-strength alloys of the 7xxx system (Al-Zn-Mg-Cu) in the powder version allow to increase the tensile strength in comparison with the cast ones by 10%—15% and reach values for the PV90 alloy (analog of cast alloy 7055): $R_m = 740$—760 MPa ; $R_{p0.2} = 650$—730 MPa, $A = 7\%$; $E = 72$ GPa[90].

In recent years, amorphous and microcrystalline alloys, characterized by high strength, heat resistance, corrosion resistance and high elastic modulus, have started to be used to meet the needs of aviation technology. The company Allied-Signal (USA) organized the production of microcrystalline alloys based on aluminum according to the following technology: production of amorphous tape by spinning (casting a metal stream on a rotating disk), grinding the tape, degassing the powder, and compacting the powder to obtain a blank for further pressure treatment. For producing aircraft parts, alloy FVS0812 (Al-8.5Fe-1.3V-1.7Si) was developed. The alloy has $\rho = 3\,020$ kg/m³, $R_m = 440$ MPa, $A = 10\%$—12%, $R_m^{315} = 280$ MPa.

Another alloy for a similar purpose FVS1212 (Al-12.4Fe-1.2V-2.3Si) has $\rho = 3\,070$ kg/m³, $R_m = 636$ MPa, $R_m^{345} = 286$ MPa, $R_m^{450} = 147$ MPa, $R_m^{480} = 126$ MPa. Metallurgical company Alcoa (USA) produces rapidly quenched aluminum alloy SU78 (Al-8Fe-4Cu) in which $R_m = 452$ MPa and $R_m^{230} = 391$ MPa. Pratt & Whitney (USA) designs an alloy Al-8Fe-2Mo, which has $R_m = 415$ MPa and $R_m^{230} = 331$ MPa. These alloys retain their strength at temperatures unattainable for standard alloys and have a chemical composition that cannot be obtained by casting.

Special attention should be paid to Al-Li alloys obtained by powder technology. The beginning of the development of Al-Li alloys was made by the member of academy I. N. Fridlander who observed the great effect of thermal hardening of the alloy system Al-Li-Mg[91]. As a result, alloy 01420 (powder analogue 01429) was developed, which had the same strength as 2024 but was 10% lighter than it and had an 8% higher elastic modulus, which made it possible to significantly reduce weight of designs (10%—15%). The increase in the lithium content in the alloy makes it possible to reduce its specific weight, to increase the elastic modulus and the effect of aging. At present, a whole range of Al-Li alloys (1424, 1430, 1450, 1460, 1470—Russia series; alloys of the 8xxx system—USA series) with a tensile strength of up to $R_m = 570$ MPa has been developed. However, they still do not have widespread industrial use in the aerospace industry due to their corrosion and fatigue properties as well as their sensitivity to "solar" heating at $T = 80$—100 °C. These properties can be improved by thermomechanical processing, on which researchers are intensively working today.

In addition to the strength properties of the powder alloys, due to the fine-grained structure, the fatigue properties are approximately 30%—40% higher, and, due to the absence of segregations, the threshold stress value during corrosion cracking increases by about 10%. On the other hand, as a result of a sharp decrease in the grain size (approximately 10 times if compared to cast alloys), a decrease in the fracture toughness is noted in powder alloys and the growth rate of fatigue cracks increases, which is a serious disadvantage.

In summary, powder metallurgy Al alloys are a special class of materials. Their feature is that in these materials the whole complex of properties is obtained not at the stage of alloy forming, but during subsequent thermomechanical processing. At the stage of production of the alloy, the possibilities to introduce damage, for example, by overheating during aging or degassing, are included. Therefore, the production of aluminum powder alloys requires strict adherence to technological discipline at all stages of production.

4.2 Powder Metallurgy Titanium Alloys

Titanium is a chemical element of the fourth group of the periodic system of Mendeleev: a) the chemical symbol is Ti; b) in solid state it has two structural modifications—① a low-temperature α-phase with a Hexagonal Close-Packed (HCP) crystal lattice having parameters $a = 0.295\ 11$ nm, $c = 0.468\ 43$ nm, $c/a = 1.590\ 7$; ② the high-temperature β-phase with a Body-Centered Cubic (BCC) crystal lattice with a period $a = 0.3282$ nm. The temperature of phase transformation α→β is $T_{pt} = 882$ ℃; atomic number is 22; atomic mass is 47.9; melting temperature $T_{mel} = 1\,680$ ℃; evaporating temperature $T_{ev} = 3\,169$ ℃; density $\rho = 4\,505$ kg/m³; elastic modulus $E = 120$ GPa; tensile strength $R_m \leqslant 240$ MPa (for titanium iodide); electrical resistivity $R = 42 \times 10^{-8}$ Ω·m; linear coefficient of thermal expansion $\alpha = 9.2 \times 10^{-6}$ K^{-1}; Poisson ratio is 0.32; thermal conductivity $\lambda = 22.065$ W/(m·K)[92].

Titanium is one of the most common naturally occurring metals (after iron, aluminum and magnesium found in the earth core) and has a unique complex of physical and mechanical properties. Even in its pure form (technical), titanium has mechanical properties at the level of thermally hardened steel. Technically pure titanium is marked in the international classification Grade 1 (Grade 2, Grade 3, Grade 4) depending on purity and has $R_m = 400$—500 MPa, $R_{p0.2} = 350$—400 MPa, $A = 20\%$, $Z = 55\%$, KCU = 98 J/cm² (high-purity iodide titanium has strength $R_m = 240$ MPa). In the Chinese classification, unalloyed titanium is marked as TA0 (TA0-1, TA1, TA2, TA3). Contamination of titanium with impurities (O_2, N_2, C) dramatically increases the strength and brittleness of the material, which is the main obstacle in its industrial application and requires precise control and strict technological discipline at all stages of manufacturing titanium parts (especially from high-strength alloys).

Alloying of titanium with other metals (first of all by Al) makes it possible to significantly increase its physical and mechanical properties, which in absolute value are comparable with the properties of heat-treated steels, including the high-strength ones[92].

In addition to high strength and plasticity with a low specific weight ($\rho = 4\,505$ kg/m³), it has excellent corrosion resistance to almost all corrosion factors of organic origin and many aggressive mineral media, which makes it indispensable in the chemical industry. Titanium alloys have high strength and resistance to oxidation up to temperature of 600 ℃. Heated to 1 000 ℃ titanium burns intensely in a stream of oxygen, if a fresh (not oxidized) scratch is applied to the surface.

Titanium alloys have a significant advantage over other structural materials in terms of specific strength in a wide temperature range from −250 ℃ to +600 ℃; low values of thermal expansion and thermal conductivity; high anti-corrosion properties; good technological properties and therefore received the name of "space age metal". However, the limiting factor for the application of titanium in industry is its high cost, low Material Utilization Ratio (MUR) using traditional production methods. In this regard, extremely great attention is paid to powder metallurgy methods for manufacturing the parts made of titanium alloys. Increasing the MUR will allow to reduce the cost of parts by 2—3 times if they are manufactured without machining. This direction is promoted by the fact that in the first stage of titanium metal production in industry, a titanium sponge is obtained (almost powder), which is then converted into a titanium semi-finished product by numerous technologies.

The technology of manufacturing titanium parts by powder metallurgy develops in two main directions: the use of alloyed powders of titanium (alloys) and the use of mixtures of powders of initial elements for the synthesis of alloys in situ[93-96].

The manufacture of semi-finished products and products from a previously obtained alloy (alloyed powders) predetermines the following technological scheme: melting of the alloy in the traditional way; spraying of a rotating slab (or rod) by plasma or electron beam heating of its end (the spraying method makes it possible to additionally alloy the metals with rare earth elements, for example, Eu or Nd, which significantly increases the heat resistance by 1.5 times); Hot Isostatic Pressing (HIP) of the obtained powder (granules) with or without further deformation. It is possible to extrude the sprayed powder in the shell at high temperatures, which is used for production of rods and profiles and allows to get a finer-grained structure than with HIP.

Obtaining powder (granules) by spraying the rotating slab (rod) allows to provide the technological methods of the suitable spherical shape of particles, whose size depends on the speed of rotation of the workpiece and has a small variation. Since spraying occurs during plasma heating in an inert gas (argon, helium, or their mixture), and during electron beam heating—in vacuum, the powder microgranules do not have oxide inclusions and impurities. The structure of such powders is micro size, consisting of several grains. Since compaction of powders occurs in hermetic vacuum shells at high temperatures, 100% density and high chemical uniformity are achieved.

This method allows to increase the homogeneity of properties and structure of the alloy, to increase the ductility, heat resistance, fatigue and corrosion properties, and to significantly reduce the production cycle and production waste. These advantages are achieved due to the formation of a uniform fine-grained structure throughout the volume of the product, the possibility of obtaining parts of complex geometric shape, which is close to the finished product. At present, semi-finished products are produced in this way from titanium alloys for structural purpose (Ti-5Al, Ti-6Al-4V, Ti-5Al-5V-5Mo-3Cr, Ti-10V-2Fe-3Al, etc.) and from heat-resistant micro-alloyed titanium alloys, used for manufacturing blades and turbines of aircraft engines (Ti-6Al-4V, Ti-6.5Al-3Mo-1.5Zr-0.25Si, Ti-6.5Al-2.5Sn-4Zr-1Nb-0.7Mo-0.15Si, Ti6.5Al-2Sn-4Zr-4Mo-1W-0.2Si). Sometimes, to obtain a more accurate form of the semi-finished products after HIP, the specialists perform hot stamping with heating to temperatures below phase transformation.

Micro-granulation technology with subsequent one or two-stage compaction provides higher strength and ductility than traditional technology[97]. The larger the sizes of the products, on which it is more difficult to ensure a homogeneous fine-grained structure and a small scatter of mechanical properties, the more significant the advantages of powder technology. Often on large, complex geometric shaped semi-finished products, it is impossible to achieve a fine-grained uniform structure using traditional methods of ingot melting and its subsequent hot deformation. Powder metallurgy allows you to solve the problem, but the disadvantage of this technology is the high cost of semi-finished products.

Much more economical is the technology using a mixture of powders of titanium and alloying elements or master alloys. In this case, the process of manufacturing of products with almost final dimensions consists of the dosage and mixing of powders, cold pressing and vacuum sintering. This technology has been used in aircraft manufacturing by various companies for a long time to manufacture non-critical low-loaded parts from Ti-6Al-4V, Ti-2Al-2Mo and Ti-8Mn alloys. However, such products are distinguished by rather high residual porosity (5%—7%) and high oxygen content, which leads to a decrease in ductility and especially fatigue properties of the alloys. It is possible to increase the density of the blanks by subsequent HIP, but this process greatly increases the cost of the final product.

A new direction in the technology of manufacturing titanium alloys from a mixture of powders is the use of titanium hydride TiH_2 instead of titanium powder. Cost-effective production technology of titanium alloys and products by the method of pressing and sintering mixtures of powder components based on TiH_2 with in-situ synthesis of the alloy without additional operations was developed, which was firstly reported by Ivasishin and the coworkers[98-99].

Titanium hydride TiH_2 is a brittle material with low strength ($R_m \leqslant 200$ MPa) and low density ($\rho = 3\,900$ kg/m^3), and its particles are easily crushed under the action of pressing, which guarantee the possibility of obtaining fairly dense and durable compacts. The phase transformation of $TiH_2 \rightarrow Ti + 2H^+$ when heated to temperatures $T = 350-500$ ℃ by the release of atomic hydrogen and lattice transformation leads to an activated state of the titanium matrix. The activation of the titanium component in the compound of powders occurs due to volume effects, accompanied by phase transformation at rather low temperatures compared to the recrystallization temperature of titanium ($T_{rec} = 750$ °C). It indicates a sharp increase in the defect density of the crystal structure, which accelerates the diffusion in the solid phase, that is, alloying and sintering of particles. In addition, atomic hydrogen, before recombination into molecules, is capable of reducing metal oxides and other impurities, which are always present on powder particles. The products obtained by reaction of hydrogen with impurities are removed from the working space of the furnace, since sintering is carried out in vacuum and this removal (including hydrogen dissolved in titanium) occurs prior to the active sintering processes of compacts. After vacuum sintering at temperature of $T = 1\,350$ °C, the residual porosity of the products is 1%—2%. Fine-grained β-transformed structure is obtained with an average grain size of about 50—150 μm. The residual hydrogen content is not more than 0.08%, which is significantly lower than the allowable.

The hydride technology was tested on products from high-strength titanium alloys with the mechanical properties presented in Table 4.2[100]. It is possible to improve the mechanical properties of high-strength titanium powder metallurgy alloys by hot plastic deformation in a two-phase (α+β) region after sintering. Such deformation changes the morphology of the grain structure of the alloy and eliminates the residual porosity. At the same time, the degree of deformation should be at least 75%. The properties of the TC4 alloy (Ti-6Al-4V) are presented in Table 4.3, and the structure before and after rolling is shown in Fig. 4.4[101].

TABLE 4.2 Mechanical properties of powdered titanium alloys

Alloys	R_m/MPa	$R_{p0.2}$/MPa	A/%	Z/%
Ti-6Al-4V	970	890	12	32
Ti-10V-2Fe-3Al	1 105	1 000	8	13
Ti-5Al-5V-5Mo-3Cr	1 205	1 140	10	14

TABLE 4.3 Mechanical properties of TC4 alloy produced by powder metallurgy technology

Alloy conditions	R_m/MPa	$R_{p0.2}$/MPa	A/%	Z/%
After sintering	970	890	12	32
After rolling	1 110	1 020	15	41

Due to the possibility of obtaining porous semi-finished products from powder blanks, the technology of asymmetric rolling of powders with subsequent sintering and repeated hot and/or cold rolling is of particular interest. Currently, researchers

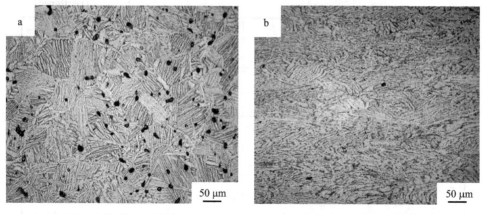

(a) After sintering and before rolling   (b) After rolling with deformation degree of 75%

FIGURE 4.4 Microstructure of the TC4 alloys produced by powder metallurgy

of titanium alloys face the task of producing alloys with strength $R_m = 1\,500 - 1\,700$ MPa and a Young's modulus $E = 200$ GPa, while preserving the plasticity at a sufficient level. One of the ways to achieve such properties is to provide ultrafine grains in the semi-finished product and to use isothermal deformation under superplasticity conditions to preserve the uniformity of the structure.

4.3 Powder Metallurgy Magnesium Alloys

Magnesium is a chemical element of the second group of the periodic system of Mendeleev: a) the chemical symbol is Mg; b) atomic number is 12; c) atomic mass is 24.305. Magnesium has a Hexagonal Close-Packed (HCP) lattice with periods $a = 3.203\,0$ Å (1 Å $= 10^{-10}$ m) and $c = 5.200\,2$ Å (at temperature of 25 ℃); the atomic radius for the coordination number 12 is 1.60 Å. Magnesium is a silvery-white paramagnetic metal. The physical and mechanical properties of magnesium are summarized in Table 4.4.

TABLE 4.4 Physical and mechanical properties of magnesium

Density (at temperature of 20 ℃)	$\rho = 1\,740$ kg/m^3
Melting temperature	$T_{mel} = 650$ ℃
Evaporating temperature	$T_{ev} = 1\,107$ ℃
Specific heat capacity (in the temperature range of 20—100 ℃)	$c = 0.246$ cal/(g · K)

(continued)

Coefficient of thermal conductivity (in the temperature range of 20—300 ℃)	$\lambda = 167$ W/(m·K)
Specific electrical resistivity (at temperature of 20 ℃)	$R = 4.47 \times 10^{-6}$ Ω·cm
Thermal coefficient of linear expansion (at temperature of 25 ℃)	$\alpha = 26 \times 10^{-6}$ K^{-1}
Tensile strength	$R_m = 120$ MPa (cast) $R_m = 200$ MPa (deformed)
Yield strength	$R_{p0.2} = 30$ MPa (cast) $R_{p0.2} = 90$ MPa (deformed)
Elongation	$A = 8\%$ (cast) $A = 12\%$ (deformed)
Hardness	HB = 30 (cast) HB = 36 (deformed)
Normal elastic modulus	$E = 45$ GPa
Shear modulus	$G = 18.2$ GPa

Magnesium alloys are the lightest material among structural materials. If using magnesium alloys for producing body parts, the mass savings can be 21%, 57% and 111% compared to aluminum, titanium alloys and steel, respectively. Its wide application is prevented by the main disadvantage—low anti-corrosion properties. This disadvantage is aggravated by such technological factor as the presence of slag inclusions in the alloy, which is very problematic to dispose when producing the parts by casting due to minor differences in the specific gravity of metal and slag. This disadvantage can be eliminated using powder metallurgy technology.

Powder metallurgy of magnesium alloys involves casting tapes with a thickness of 20—30 μm using high-speed crystallization ($V_{cr} = 10^6$ ℃/s), its subsequent crushing and obtaining semi-finished products.

The advantages of powder technology based on the mechanical properties of magnesium alloys are presented in Table 4.5. In addition to mechanical properties, powder alloys have a higher corrosion resistance.

TABLE 4.5 Properties of magnesium alloy produced by casting (MA2-1 and MA14) and powder technology (MA2-1P and MA14P)

Alloy	R_m/MPa	$R_{p0.2}$/MPa	A/%	KCU/(kJ·m^{-2})
MA2-1	260	150	8	80
MA2-1P	310	230	14	600
MA14	320	150	7	90
MA14P	340	270	13	300

4.4 Powder Metallurgy Beryllium Alloys

Beryllium is a chemical element of the second group of the periodic system of Mendeleev; a) the chemical symbol is Be; b) in the solid state it has two structural modifications—① a low-temperature α-phase with a Hexagonal Close-Packed (HCP) crystal lattice having parameters $a=0.228\,55$ nm, $c=0.358\,33$ nm, $c/a=1.567\,8$; ② a high-temperature β-phase with a BCC crystal lattice. The temperature of phase transformation α→β is $T_{pt}=1\,277$ ℃; atomic number is 4; atomic mass is 9.013; melting temperature is $T_{mel}=1\,285$ ℃; evaporating temperature $T_{ev}=2\,450$ ℃; density $\rho=1\,848$ kg/m³; elastic modulus $E=300$ GPa; heat capacity $c=1.84$ kJ/(kg·℃)[90].

The use of beryllium in aviation, rocket and space technology is determined by a combination of its low density ($\rho=1\,848$ kg/m³), high melting temperature ($T_{mel}=1\,285$ ℃), an unusually high elastic modulus ($E=300$ GPa), corrosion resistance in atmospheric conditions and specific nuclear characteristics (low thermal neutron capture cross-section is 0.009×10^{-28} m², high neutron scattering cross-section is 6.9×10^{-28} m², and the ability to participate in (α,n) and (γ,n) reactions). Beryllium is characterized by the highest specific heat capacity among solid metals ($c=1.84$ kJ/(kg·℃)). The low density and high modulus makes it very effective in the construction of aircraft operating on buckling.

At present, several technological schemes have been developed for obtaining sintered beryllium products:

① cold pressing at pressures of 500—800 MPa, followed by sintering in vacuum at $T=1\,050$—$1\,100$ ℃ for 2—3 h;

② hot pressing at pressure of 800 MPa and temperatures of 550—600 ℃ or pressures of 10—15 MPa and temperatures of $1\,000$—$1\,100$ ℃ in an inert gas (argon) environment;

③ cold pressing, sintering and hot stamping at temperatures of 600—$1\,000$ ℃

at the air of pre-sintered blanks that using induction heating allows to obtain 100% density;

④ cold pressing sintering and hot extrusion;

⑤ impregnation of pre-sintered blanks.

Production of beryllium products by powder metallurgy allows to achieve significantly higher physical and mechanical properties than those produced by casting in vacuum. Beryllium obtained by casting in vacuum has $R_m=140$ MPa with $A=0.1\%$. Hot pressing at temperature of 1 000 ℃ and a pressure of 20 MPa allows to obtain metal with $R_m=478$ MPa and $A=1.4\%$[90].

The disadvantages of beryllium are low strength properties, limited application at temperatures above 540 ℃, as well as difficulties with environmental protection due to toxicity of beryllium oxides. To increase the operating temperature of beryllium alloys, the method of rapid crystallization is used, which ensures the formation of dispersion phases.

Due to its lightness, high dimensional stability and the possibility of precision processing, beryllium is an ideal material for use in navigation devices such as gyroscopes. Beryllium is used in space optics for producing scanning mirrors. Polished metal has an infrared reflectivity of more than 95%. Beryllium wire with a diameter of 0.1—0.5 mm can be used as an effective reinforcement for composite materials. The nuclear characteristics of beryllium determine its use in various kinds of special equipment, in particular, for producing the entrance and exit windows in radiation detectors, electrophysical equipment, controls at airports, defectoscopes and other instruments used in aviation technology. For these purposes, thin vacuum-tight foils are used, which allow to ensure the ecological and corrosion safety of products. Beryllium foil is a very high-tech and expensive product.

Beryllium is a promising structural material for reusable spacecraft, in particular for manufacturing the landing gear brakes. Given the high landing speeds and a large mass of reusable spacecraft, high-performance braking mechanisms are needed, for the manufacture of which beryllium is used as the force framework of the brake and the carbon-carbon composite—as friction material. Only under these conditions it became possible to satisfy the requirements for the weight limit of the brake wheel.

Beryllium brake grade blanks in the form of discs are produced by cold pressing of powder and hot compaction at temperature of 1 000—1 200 ℃, with subsequent deformation of 70—80% at temperature of 1 000—1 100 ℃. A promising direction for using the beryllium in the form of sheets is protection of the design of spacecraft from micrometeorites.

However, all the advantages of beryllium are crossed out by one disadvantage—it has a detrimental effect on the human body if it gets inside, causing an incurable berylliosis disease associated with softening of human bone tissue. Therefore, its use is limited to those areas where it is indispensable.

Chapter 5 Heat Resistant Materials and Materials Strong at High Temperatures

The development of aircraft engine technology, rocket technology and astronautics is largely determined by the use of heat resistant materials and materials that are strong at high-temperature. For parts of the hot section of aircraft and rocket engines, external shell of space rockets, the properties of heat resistance and high-temperature strength are decisive for development of the construction and further operation.

Heat resistance is the ability of materials to resist chemical destruction of the surface under the influence of hot air or other gases, more often the products of fuel combustion. There are ceramic (based on refractory oxides and their compounds or based on oxygen-free refractory compounds—carbides, borides, nitrides, silicides), ceramic-metal, metal, carbon and graphite, glass-ceramic and vitro-crystalline, heat resistant composite materials. Thermomechanical properties of heat resistant materials are improved by the introduction of reinforcements (particles, fibers) of dispersion-strengthening phases, as well as by prevention of collective recrystallization at high temperatures[102]. An indirect assessment of heat resistance can be the resistance of materials to oxidation under given temperature conditions, determined by the change in sample mass per unit surface.

High-temperature strength is the property of a material to resist deformation and destruction under the influence of mechanical loads at high temperature. High-temperature strength is a complex concept which includes the following high temperature properties:

- *Creep resistance* is the ability of a material to resist slow plastic deformation during continuously applied loads (below the yield strength of the material) at elevated temperatures. The creep strength is marked as $R_{\varepsilon/t}^{T}$, where T is the test temperature, ℃; ε is the specified total or residual strain, %; t is the duration of the test, h. When determined by the creep rate, the creep strength is marked as R_{v}^{T}, where T is the test temperature, v is the given creep rate.
- *Short-time strength* is determined by tests on tensile at a given temperature R_{m}^{T}, where T is the test temperature, ℃.
- *Stress-rupture strength* is the ability of a material to resist destruction

during long-term influence of static load and high temperature. It is characterized by the limit of stress-rupture strength—the highest stress causing the destruction of the material within a certain time at a constant temperature R_t^T, where T is the test temperature, ℃; t is the given test duration, h.
- *Long-term ductility* characterizes the deformation (ε_p, %) to failure at a given temperature and load.

The advantage for materials that are strong at high temperature is the stability of the mechanical properties. For most alloys that are strong at high temperature, it is impossible to achieve full stabilization of the properties, and as a result, the acceptable limits for these properties over time are set.

Materials that are strong at high temperature include alloys based on Fe, Ni, Co, refractory materials (metal oxide, metal carbide and intermetallic compounds). A variety of alloys that are strong at high temperature are composite materials consisting of heat resistant matrix (alloys based on Fe, Ni, Co; systems of Fe-Ni-Cr alloys, Ni-Cr-metal, Co-Cr-metal and others) reinforced with refractory metals and their alloys or oxides in the form of fibers, threads, particles.

5.1 Refractory Metals and Alloys

Refractory metal is a conventional term that defines that this metal has a melting point of at least 2 000 ℃ (sometimes it is stated as 1 800 ℃) and is used in industry. Such metals are traditionally produced by powder metallurgy, for example, W, Mo, Nb, Ta, Zr, V, Cr, Hf, Re[103-104].

In aerospace technology, the "big four" refractory metals are used the most frequently—W, Ta, Nb and Mo, whose melting points are $T_{mel(W)} = 3\ 380$ ℃, $T_{mel(Ta)} = 2\ 996$ ℃, $T_{mel(Nb)} = 2\ 468$ ℃, $T_{mel(Mo)} = 2\ 610$ ℃, respectively. All metals of the "big four" are actively oxidized in air at relatively low temperatures, and Ta and Nb are also prone to gas saturation, therefore, their effective use can occur in absence of the "gas - metal" reaction. In case of absence of the oxidizing environment, creep and collective recrystallization, which limit the service life of a high-temperature node, are the main causes of the malfunction of parts made from refractory metals. To reduce the influence of these factors, additives are introduced into the refractory metals (W is introduced into Mo, and Re or Th is introduced into W[59]).

Considering that the melting points of the refractory metals are higher than the operating temperature of most fire-resistant materials, it is very difficult to produce them by the smelting method. Therefore, standard powder metallurgy methods are used—pressing and sintering (or HIP), followed by hot deformation to obtain the necessary semi-finished products. The main refractory metals and their alloys, as well as the fields of application of such materials are presented in Table 5.1[105].

TABLE 5.1 The Application of refractory materials

Material	Application area
Refractory metals Mo Ta W Nb	Manufacturing of parts of thermally stressed nods of high-speed aircraft, rockets and spacecraft apparatus—heat shields, ionizers in ion rocket engines, nozzles for rocket engines, blades, fringing, body
Mo based alloys Mo-0.5Ti Mo-0.5Ti-0.08Zr Mo-25W Mo-50W	Parts operating in conditions where a combination of high hardness and strength at elevated temperatures is necessary—rocket nozzles, ionizers in ion rocket engines
Ta based alloys Ta-10W Ta-10W-1Zr Ta-30Nb-7.5V Ta-8W-2Hf	Parts of rocket engines, combustion chambers for control nozzles, coating the front edges of a wing
W based alloys W-(0.5—2.0)ThO$_2$ W-Ag	Parts operating up to temperature of 1 900 ℃, rocket nozzles operating on solid fuel
Nb based alloys Nb-10W-1Zr-0.1C Nb-30Ta-0.8Zr Nb-20W-1Zr Nb-Hf-W	Parts operating in conditions that require high strength characteristics at elevated temperatures, expansion chambers for control nozzles

When manufacturing products from refractory metals and alloys based thereon, HIP is most often used at temperature of $0.5T_{mel}$ — $0.6T_{mel}$ in gas stoves, where such inert gases as Ar or He most often serve as the working medium. This technology allows to obtain almost dense material.

Tungsten and molybdenum do not have phase transformations in the solid state, therefore their elastic modulus and tensile strength decrease monotonically with increasing of temperature, which is an advantage to some extent. High purity deformed annealed tungsten has $R_m^0 = 200$ MPa and $R_m^{2\,000} = 100$ MPa, for molybdenum $R_m^{1\,400} = 200$ MPa[105].

High-temperature strength in long-term operation is limited by creep, which in tungsten (depending on the structure and degree of purity) is observed at temperature of about 1 600 ℃, and in pure molybdenum at 850 ℃. High-temperature strength and fatigue properties are determined by recrystallization processes, which take place actively in tungsten at temperature of 1 200—1 300 ℃, and in molybdenum at 1 000 ℃[105].

Tungsten as a structural material is used to produce solid-propellant, ion electrojet rocket engines. Since rocket nozzles operate in reducing atmosphere or vacuum, pure tungsten (or molybdenum) can be used to manufacture them without fear of oxidation at high temperatures. In constructions of rocket engines, it is advisable to use thin-walled tungsten inserts with a substrate that intensively removes heat. The rotational extrusion technology makes it possible to manufacture thin-walled rotating parts from powder metallurgy tungsten sheet of 2—25 mm thickness with a diameter of more than 1 meter. In this way, the output conical nozzles and throat inserts of rocket engines, engine facing with a thickness of 10 mm are made. For parts of tungsten and molybdenum rocket engines of a more complex form, stamping is used in closed dies.

Alloyed molybdenum with Ti, Zr, V, Nb and other elements that increase its heat resistance, is used for manufacturing components of aviation gas turbines and rocket engines (working and nozzle turbine blades, exhaust nozzles and combustion chambers of direct-flow rocket engines, front edges of rockets, nasal cones and so on). The service lifetime of rocket engines is so small that the oxidation of molybdenum does not matter, but the important thing is the high-temperature strength. The working and nozzle blades of gas turbines require protection against oxidation, which is carried out by heat-resistant coatings.

To obtain TZM (Mo-0.5Ti-0.08Zr-0.015C) alloy (all compositions expressed in wt.% unless otherwise noted), molybdenum powder is mixed with hydrides TiH_2 and ZrH_2 and carbon black. The powder mixture is pressed at pressure of $P = 200$—400 MPa. When heated in vacuum for sintering, hydrides decompose with the release of atomic hydrogen, which helps clean the surfaces of the powder particles from oxides, which significantly improve the ductility of the alloy. The alloy elements dissolve in molybdenum, and when cooled after sintering, they precipitate as finely dispersed carbides due to a decrease in the solubility. These alloying elements significantly increase the recrystallization temperature of molybdenum and improve high-temperature strength and long-term strength as a result of highly dispersed carbide phases. The addition of Nb to such alloys further enhances the long-term strength[105].

Experimental turbine rotors made of forged molybdenum alloy TZM with a silicide coating successfully worked in a reactive experimental engine of Williams International for 7 h at a temperature of 1 343 ℃, which indicates the possibility of using molybdenum alloys in short-life military engines.

For aviation and rocket and space technology, tungsten and molybdenum alloys with hafnium, boron and carbon (W-0.5Hf-0.021C), (Mo-0.8Hf-0.3B-0.03C) are promising, with $R_m^{1\,500}=600—700$ MPa and having good technological properties—workability and weldability[105].

Alloys of tantalum with tungsten, hafnium and boron are among the materials that are strong to high-temperature for rocket-space purpose[105]. Tungsten and hafnium contribute to an increase in the elastic modulus of the alloys, melting and recrystallization temperatures. The Ta-10W alloy which is more plastic than tungsten, does not inferior to it in strength, and guarantee oxidation resistance at temperatures up to 2 800 ℃. the Ta-8W-2Hf alloy has the high specific strength at elevated temperatures. The tantalum based alloys are used for manufacturing combustion chambers and nozzles of jet engines, the front edges of the tail of supersonic aircraft[106-107].

When producing tantalum alloys by powder metallurgy, the mixture of powders is pressed under pressure $P=210—850$ MPa, at which a density of 0.6—0.7 of the theoretical value is reached, and then subjected to sintering at temperature of $T=1\,980—2\,500$ ℃ for several hours in vacuum. To obtain almost dense material, after sintering, forging or rolling is performed.

Niobium can partially or completely replace the difficult-to-obtain tantalum with a significant economic effect (it is cheaper and almost twice lighter than tantalum). Niobium alloys are characterized by high heat resistance and relatively low density[108].

Structural alloys of niobium and tantalum have found application in liquid and solid-propellant rocket engines. In this case, insufficient oxidation resistance is not particularly important due to the short duration of operation of such engines.

Niobium alloys are used in the gas turbine engine for military purposes—the flaps of an adjustable nozzle are made of Nb-10Hf-1Ti alloy with silicone coating. The alloy has $T_{mel}=2\,349$ ℃ and $\rho=8\,859$ kg/m³, which makes it perfect for use in aviation and rocket and space technology. The Metal Technology company reported that they managed to obtain parts from the specified alloy by three-dimensional printing[109].

Chromium is the only one of all refractory metals that does not need to be protected from gas corrosion at high temperatures. The maximum temperature of long-term operation of chromium and its alloys without the use of protective coatings is within 1 200—1 350 ℃. The ability to operate such parts without protection against oxidation is due to the formation of a dense and durable oxide film that protects the entire product from further oxidation at high temperatures[110].

Sintered chromium alloys are made by cold pressing of a mixture of powders at pressure of 300—400 MPa, followed by sintering in hydrogen at temperature of 1 450—1 500 ℃[111].

5.2 Superalloys

Superalloys are alloys based, as a rule, on elements of Group Ⅷ of the periodic system of chemical elements of Mendeleev designed for operation at elevated temperatures under the influence of relatively high mechanical loads and conditions where materials are often required to have high surface stability. Depending on the base material, there are three main classes of superalloys—nickel, cobalt and iron-based superalloys. In addition, a subgroup of iron-nickel superalloys is distinguished, containing significant amounts of Ni and Fe and possessing metallurgical properties similar to nickel superalloys.

Superalloys are operated at temperatures close to the melting point—these are parts of the hot path of the gas turbine engines of airplanes, spacecraft, and rockets. Production of jet engines leads to the necessity to produce superalloys.

Superalloys, as a rule, have low plasticity and are difficult to handle by mechanical process (deformable superalloys) or cannot be deformed at all (casting superalloys). Powder metallurgy creates new opportunities in the technology of superalloys, as it allows to obtain large-sized parts (turbine disks, rotor blades, etc.) with a uniform structure and high properties even from non-deformable superalloys. The increased physical and mechanical properties of powder superalloys are provided by the structural factor (fine grain structure) and the absence of segregations of alloying elements, which is unavoidable during production by the casting method.

The potential of powder metallurgy in the production of superalloys was shown at the end of the 60s of the 20th century, which, in the 70s, led to creation of a gas turbine, which was almost completely manufactured by powder metallurgy methods, including the components—rotor blades, discs and other important structural components[112].

Superalloys made by powder metallurgy are used in cases, where the standard approach—casting or stamping does not meet the requirements of a particular design. For example, IN—100 alloy manufactured by standard technology has the mechanical properties: $R_m = 1\ 054$ MPa, $A = 8\%$, and a powder alloy with the same chemical composition is $R_m = 1\ 406$ MPa, $A = 12\%$. Since superalloys are complex alloyed alloys with a high content of alloying elements, it is usually difficult to avoid segregation when casting is used, especially for massive blanks. The presence of segregation in most cases of the use of cast products is the main reason of destruction of parts in operation[112].

The advantages of powder metallurgy technology for producing superalloy are:

① reduced tendency to segregations, which contributes to the formation of smaller intermetallic precipitates and reduces the distance between dendrites to values that cannot be obtained when manufacturing the usual size by casting;

② susceptibility of the microstructure to deformation, which allows to obtain materials with very high mechanical properties ("microscale" homogeneity of compacted powder ensures obtaining material properties of extremely high uniformity and reproducibility);

③ the possibility of producing materials with a unique structure for works in special conditions (for example, dispersion-strengthened superalloys cannot be obtained by any other methods except powder metallurgy)[112].

The main disadvantages of powder technology superalloys are:

① the main properties that determine the possibility of long-term operation under high temperature and loads are fracture resistance and fatigue. Due to the presence of defects in the form of pores or impurities in the powder alloys, the durability properties decrease.

② the necessity of complex thermomechanical processing of particularly important parts in order to neutralize the effect of defects associated with impurities.

③ the necessity to use a very fine powder (which reduces productivity) to regulate the maximum size of defects.

④ the necessity for accurate and strict implementation of all technological operations for the manufacture of the finished product, including the modes of thermomechanical processing.

⑤ high (higher than the traditional technology) cost of products, due to the necessity to perform costly technological operations to ensure the integrity of the product.

⑥ the limited capabilities of non-destructive testing methods.

The main methods for producing superalloy powders are the atomization of the prepared alloy with an inert or dissolved gas. In the first case, the metal of vacuum cleaning is supplied to the ladle, where, at a strictly specified temperature, it is discharged through a calibrated orifice. The metal jet is crushed by an inert gas, which is fed under high pressure to the metal jet. When this occurs, the jet of metal is crushed into spherical droplets and a cooling rate of ~100 ℃/s is reached[112].

In the second case, the medium in which melting takes place (under overpressure) contains some amount of gas dissolved in the metal (usually hydrogen). Atomization is carried out by immersing a tube (usually ceramic) into the melt. The other end of the tube is brought into the upper vacuum spray chamber. Liquid metal under pressure of the working medium rises up the tube and into the chamber, where it is sprayed, caused by a sharp pressure drop and rapid release of gas dissolved in the melt. The cooling rate of powder particles in this method is ~1 000 ℃/s.

For producing superalloy powders, Pratt & Whithey Aircraft company first proposed a centrifugal atomization method. During this process, a stream of metal melted in an induction furnace is dosed to the surface of a rapidly rotating disk. The layer of liquid metal is accelerated at the periphery of the disk, where occurs the formation of spherical particles-droplets, which are sprayed with a vertical jet of inert helium gas. As a result of secondary atomization, the cooling of the particles occurs at a very high speed. The powder obtained in this way is characterized by very narrow particle size distribution. After atomization, powders are stored in an inert medium until compacted.

Powder technology of production of superalloys consists in consolidating the atomized powder using the Hot Isostatic Pressing (HIP) method or hot pressing followed by extrusion. Sometimes after HIP, hot extrusion or isothermal stamping is performed, which optimizes the structure of the material and improves the mechanical properties.

With HIP, the powder placed in the container is heated after evacuation to an elevated temperature (above or below the dissolution temperature of the γ'-phase—stable intermetallic phase of Ni_3Al—depending on the required grain size in the finished product) and is maintained at the selected temperature below gas pressure ~100 MPa. Such exposure of temperature and pressure results in dense structure from the powder.

The main goal of HIP is to obtain an absolutely dense material with a structure without any traces of the primary boundaries between the original powder particles. To ensure a non-porous state, a rather long exposure time at a certain temperature and pressure is required. By varying these parameters, one can change the dominant compaction mechanism—either the rapid plastic deformation of powder particles or the creep compaction that occurs more slowly. By changing the compaction mechanism, it is possible, to some extent, to influence the final material structure of the finished product. Some parts after HIP are heat-treated with very strict compliance with its regimes in order to avoid their cracking.

Another technology—hot pressing of the powder at a temperature below the temperature of the dissolution of the γ'-phase, followed by extrusion, is a more common method of consolidation. As in the first case, the powder is placed in a container, vacuumized, heated and subjected to hot pressing in closed dies. After pressing, hot extrusion is carried out with a ratio of at least 6 : 1 to obtain completely dense material. Such material has a recrystallized fine-grained structure in which the primary boundaries between the powder grains and any traces of the dendritic structure of the original powder are almost completely absent. The obtained structure depends on the particle size of the original powder and the technology for its preparation.

Thermomechanical processing of powder metallurgy superalloys is carried out to:

① neutralize the effects of defects associated with pollution. These can be internal defects that were originally present in the original powder and are associated with the processes of its manufacturing (ceramic inclusions, pores filled with argon, oxidized particles) or external defects arising from further processing of the powder.

② reduce the cost of parts due to the rational use of the material in their production. Due to their superplasticity, it is possible to obtain high-precision forgings that are as close as possible to the finished parts.

③ achieve the required quality of metal due to microstructural changes associated with the plastic flow of grains, the formation of bimodal structure.

As a rule, powder metallurgy superalloys made without violation of technology have good mechanical properties and fairly high uniformity. Forging leads to only a slight improvement in static properties. At the same time, thermomechanical processing leads to an increase in the minimum level of dynamic properties, whose values are determined by the content of defects in the material.

Powder metallurgy superalloys can be forged in several ways; however, it is most appropriate to use isothermal deformation when the tooling and the deformable material have the same temperature. Thus, the powder of Rene95 alloy consolidated at temperature below the solvus (the solubility temperature of the γ'-phase) (see Table 5.2) has a fine-grained (2—5 μm) structure and has superplastic properties, that is, at low strain rates, the flow of this material occurs at very low tensions. The deformation results in an increase in stress to a peak value and the subsequent softening of the flow of material with a constant rate of deformation accumulation[112].

TABLE 5.2 Chemical compositions of powder metallurgy superalloys

Elements	Alloy series							
	Rene95	IN100	MERL76	APK1	NiAlMo	EP741	EP962	EP975
Ni	Bal.	Bal.	Bal.	Bal.	Bal.	Bal.	Bal.	Bal.
Cr	13.5	12.4	12.5	15.5	0.1	9.0	13.5	9.0
Co	8.0	18.5	18.5	17.0	—	15.5	9.0	11.0
Mo	3.5	3.2	3.2	5.0	12.0	3.8	4.5	2.8
W	3.5	—	—	—	8.0	5.5	2.7	9.5
Nb	3.5	—	1.4	—	—	2.6	3.5	2.0
Al	3.5	5.0	5.0	4.0	7.8	5.0	3.8	5.2
Ti	2.5	4.3	4.4	3.5	—	1.8	2.7	2.7
Hf	—	—	0.4	—	—	0.3	0.2	0.2
Ta	—	—	—	—	0.6	—	—	—
V	—	0.8	—	—	—	—	0.4	—
Zr	0.05	0.06	0.06	—	—	—	—	—
B	0.01	0.02	0.2	0.02	—	0.02	0.01	0.02
C	0.05	0.07	0.04	0.03	—	0.06	0.06	0.06

In general, forging is carried out at a rate of 1 mm/min at temperature of 1 200 ℃ and a flow tension of 50—100 MPa, depending on the strain rate. Hot stamping forging can also be carried out at high strain rates (up to several min^{-1}), however, the yield strength of the finished product and tool lifetime are deteriorated. After forging, such alloys are usually subjected to partial annealing (at temperatures below the solvus) with one or more aging cycles.

The APK1 material (see Table 5.2), which is a low-carbon modification of the Astroloy powder alloy, is made by the HIP method at temperatures above the solvus line, which is necessary to obtain a fairly coarse-grained structure.

Thermomechanical processing of the alloy is carried out on a standard forging equipment at high deformation rates (up to 1 000 mm/min). As a result, a so-called "necklace" microstructure is formed, a characteristic feature of which is the presence of large deformed primary grains surrounded by a necklace of small recrystallized grains. This structure provides a good combination of the reduction in the nucleation rate and the rate of crack propagation. Attempts have been made to create such structure in the Rene95 powder alloy, however, in this case, preference was given to a more noble fully recrystallized fine-grained structure[112].

Alloys from powder of NiAlMo series (see Table 5.2) obtained by centrifugal spraying using the Pratt & Whitney Aircraft technology are used to make gas turbine blades. The consolidation of powder alloys of NiAlMo series is carried out by the same methods as described above and the subsequent thermomechanical treatment is carried out according to such conditions in order to obtain a highly textured structure with elongated grains. These materials are not inferior in their properties and sometimes even surpass the cast alloys used to manufacture the blades of the most advanced gas turbines.

The chemical compositions of the main powder metallurgy superalloys are given in Table 5.2. The static mechanical properties of powder metallurgy superalloys (see Table 5.3) depend on their chemical compositions and structures, which in turn are determined by the size of the powder particles, the method of their consolidation and the heat treatment regime[112].

TABLE 5.3 Mechanical properties of powder superalloys

Alloys	Tensile tests					Durability		
	$T/℃$	R_m/MPa	$R_{p0.2}/MPa$	$A/\%$	$Z/\%$	$T/℃$	Stress/MPa	Average durability/h
U700*	760	1 027	1 020	20	28	760	586	25
IN100**	704	1 268	1 065	20	21	760	655	35
Rene95***	649	1 503	1 138	13.5	14.9	649	1 034	54
Astroloy (low carbon)	649	1 324	979	25.6	25.9	760	634	89

Note: * HIP+heat treatment;
** Prepared from powder fraction—100 mesh, extrusion+isothermal forging;
*** Prepared from powder fraction—500 mesh, HIP at temperature of 1 120 ℃+heat treatment.

Manufacturing disks for engines of the latest generation is a difficult and highly demanding task. The solution to this problem is somehow facilitated by the relatively lower temperatures of the disks compared to the blades, but the manufacturing of the engine disks is also complicated due to the much higher requirements for reliability and complexity of the geometric shape. The most suitable materials for turbine disks are nickel alloys, which are capable of ensuring the efficiency and reliability of aircraft engines with a service lifetime of 20—50 thousand hours.

Powder metallurgy superalloys are used to make disks for compressors and turbines, as well as other parts operating at temperature of 540—670 ℃[112]. Disks for compressors weighing up to 100 kg from astroloy and IN100 alloys are produced by forging from extruded rods. Such alloys are superior to standard alloys in grain size, structural homogeneity, and lack of segregation of elements. In the extrusion process at temperature of 1 050 ℃, dynamic recrystallization occurs, but the grain size remains quite small (~ 0.5 μm).

EP741, EP962 and EP975 powder metallurgy alloys (see Table 5.2) are used for producing gas turbine engine disks. Since the granules of the sprayed alloy are characterized by a high specific surface, the bulk of non-metallic inclusions (Al_2O_3, TiN) are formed on the surface of the particles, which can be mechanically removed, and this creates a unique opportunity to produce ultrapure metal by inclusions. Ceramics are separated from metal in a jet mill in an argon jet, followed by separation of metal and ceramics in electrostatic multistage separators. Thermally degassed in vacuum, the granules are compacted by the HIP method, hot isostatic stamping or extrusion. To obtain high strength and ductility properties of powder alloys, thermomechanical processing is used, which gives the required structure.

When using granular technology, the laboriousness of manufacturing disks and consumption of metal are reduced in 1.5—2.0 and 2—3 times, respectively, while their cost is 15%—20% lower than with traditional technology. At the same time, the homogeneity of the chemical composition and properties is much higher, which significantly increases the reliability of operation. The long-term strength of powder alloys is 25%—45% higher than that of the material obtained by the traditional way, which makes it possible to raise the working temperature by 30—50 ℃, reduce the engine weight by 10%—25% and increase the service lifetime by 5—10 times[113].

Among the powder metallurgy nickel alloys that are strong at high temperatures, there should be allocated a group of alloys that combines the properties of structural and functional materials. An example is the alloy of the Ni-Cr-Mo-Al-Co-Nb-Hf doping system obtained by powder metallurgy. The use of such alloy makes it possible to principally abandon the bolted joints of the disks in the drum of a high-pressure compressor of an aviation Gas Turbine Engine (GTE), which improves the weight efficiency and reliability of the engine.

One of the effective ways to improve the mechanical properties of powder alloys that are strong at high-temperature is to develop technologies that increase the cooling rate of particles during crystallization (up to 10^6 ℃/s and higher). When ultrafast crystallization occurs, a transition of the dendritic structure into an equiaxial fine-crystalline structure is observed, segregation processes in the crystallizing metal decrease or disappear, and higher homogeneity and homogeneity of high-alloyed nickel alloys are obtained up to the disappearance of the eutectic structure. Such crystallization conditions contribute to a significant increase in solidus temperature, which is an important factor for increasing the high-temperature properties of standard alloys and enables to produce new alloys.

The increased solubility of the components associated with higher cooling rates makes it possible to move to a new class of alloys with a high level of doping, which makes it possible to reach the content of the strengthening γ'-phase up to 80%. Such alloys have a melting point, as a rule, in 40—85 ℃ higher than that of standard nickel alloys. An increase in the melting temperature expands the homogeneity range of such alloys and generates additional conditions for their heat treatment and increase in performance properties. Alloys obtained by ultrafast crystallization have unique capabilities. Due to their fine-grained structure, after HIP they differ in a compact state by improved ductility or superplasticity. Using these blanks, you can make complex stamping similar in shape to the finished parts or rolling sheets.

5.3 Dispersion-strengthened Material

An effective way to increase the high-temperature strength of metals and alloys is the introduction of inert reinforcing phases into the alloy, which is possible only with the help of powder metallurgy methods. Without interacting with the matrix at temperatures below T_{mel}, incoherent particles allow, in addition to the known methods of "direct" hardening (solid solution, coherent precipitates, etc), to implement the mechanism of "indirect" hardening, affecting the shape and texture of grains, and the shape of intergranular boundaries as a result of thermomechanical processing. The relative role of "indirect" hardening increases dramatically at temperatures of $0.7T_{mel}$—$0.8T_{mel}$, and at higher temperatures, this mechanism becomes predominant. As a result, the highest operation temperature of such alloys can exceed the operation temperature of classical alloys by 100—200 ℃, which is quite important for modern engines.

Metallic materials hardened by particles of refractory compounds—dispersed inclusions, are called *dispersion-strengthened materials*. The maximum hardening effect is achieved with a sufficiently small particle size of the hardening phase (0.01—0.05 μm), their uniform distribution in the structure of the material and the optimum distance between the particles (0.1—0.5 μm). The total volume content of particles usually does not exceed 5%—10%. Dispersion-strengthened materials are not fundamentally different from aging alloys. However, unlike the classic aging alloys, in which the strengthening dispersed phase is separated from the supersaturated solid solution, this phase is artificially introduced into dispersion-strengthened alloys at one of the stages of their manufacture. Oxides (Al_2O_3, SiO_2, ThO_2, HfO_2, ZrO_2, Cr_2O_3 and others) carbides, nitrides and carbonitrides of transition metals, intermetallic compounds such as Ni_3Al, $MnAl_6$, some refractory metals are used as hardening phases[59].

To ensure a long hardening effect, especially at elevated temperatures, the particles of the hardening phase should not interact with the matrix metal, be thermodynamically stable in the working temperature range, and should not tend to coalescence at high temperatures. In this case, high temperature strength of the material is provided, which is higher than when the main material is added with other elements.

Dispersion-strengthened materials are characterized by a continuous increase in the ratio of their strength to the strength of the matrix metal with increasing temperatures. This is explained by the fact that with the presence of a second phase in the matrix, the dislocation slip in the metal grains becomes possible at significantly higher tensions, the boundaries of these grains are blocked, and the grain growth is slowed down almost to the melting point of the matrix. The effect of grain boundary hardening is summed with the hardening itself from the particles, and the thermal stability of the structure contributes to the preservation of grain boundary hardening to the premelting temperature.

The interest of dispersion-strengthened materials to be used in rocket and space technology is determined by the stability of their properties at elevated temperatures. The first dispersion-strengthened material used in rocket and space technology was thoriated tungsten. The W-2ThO_2 alloy has $R_m^{2\,760} = 80$ MPa for wire and $R_{100}^{2\,200} < 10$ MPa for sheets[59].

To ensure a uniform distribution of thorium oxide, thorium nitrate $Th(NO_3)_4$ is introduced into the water pulp of WO_3. After evaporation, drying, calcining and reduction with hydrogen, tungsten powders with an additive ThO_2 are obtained, from which the rods are made by pressing and sintering, which are then subjected to thermomechanical processing (forging, broaching, annealing) to obtain the semi-finished product. The W-5Re-2ThO_2 alloy (rhenium is introduced to increase the strength and ductility of the alloy at low temperatures) has $R_m^{1\,925} = 110$ MPa[59].

In beryllium alloys, BeO or Be_2C is used for dispersion hardening. At temperature of 600 ℃, the Be-0.8BeO_2 alloy has $R_{p0.2} = 115$ MPa, and the Be-3BeO_2 alloy has $R_{p0.2} = 175$ MPa. High creep resistance is achieved by hardening the Be_2C alloy. With a content of 2.5% Be_2C, the value of R_{100}^{65} and R_{100}^{730} increases by 3 and 5 times, respectively, compared to pure beryllium[59]. The main application area of beryllium dispersion-strengthened alloys is rocket production[111].

The hardening phase in magnesium alloys is MgO. To maintain good processability, the oxide content is maintained at 0.1%—1.1%. The Mg-1MgO alloy has $R_{100}^{500} = 10$ MPa (the temperature of 500 ℃ is 0.85T_{mel}). Additional introduction of beryllium into the matrix material (up to 8%) with the aim of increasing heat resistance allows the use of magnesium dispersion-strengthened alloys in rocket production[111].

The best hardening phase for chromium dispersion-hardened alloys is MgO, the presence of which shifts the brittleness threshold to negative temperatures. The Cr-6MgO-0.5Ti alloy has $R_{100}^{1\,370} = 33$ MPa. Dispersion-hardened chromium-based alloys are characterized by high heat resistance and erosion resistance.

In zirconium materials, the hardening phases are ThO_2 and Y_2O_3. At temperature of 650 ℃, the ultimate strength of such material is 2—3 times higher than pure zirconium. At the same time, the creep rate is significantly reduced. Zr-Zr-5Y_2O_3 alloy has $R_{100}^{550} = 85$ MPa[59, 111].

The first dispersion-strengthened nickel alloy with a content of 2% ThO_2 was developed in the 60s years of the 20th century. Currently, nickel alloys, dispersion-strengthened by thorium oxide are manufactured by Du Pont (USA) and Cherry Gordon (Canada) under the respective trademarks TD-Ni and DS-Ni and are used in rocket technology.

The use of ThO_2 as a hardening phase is due to its slight tendency to aging, high melting point and the ability to form stable salts with particle sizes of 10^{-5}—10^{-4} mm that are optimal for the hardening phases. The powder is obtained from aqueous solutions of nickel salts, which include the ThO_2 sol, or insoluble hydroxide or nickel oxalate. ThO_2 particles serve as nuclei of crystallization. The precipitate is dried, the oxide is calcined, and the resulting powder containing ThO_2 particles is crushed, then pressed, sintered, and thermomechanically processed to produce sheets or rods. The dispersion-strengthened Ni-2ThO_2 alloy is forged in a wide temperature range, from 60 ℃ to 1 200 ℃[114]. The material is not sensitive to overheating and notching. It is well processed by cutting, has high resistance to shock and cyclic loads, is corrosion-resistant, and is not subject to intergranular and high-temperature oxidation.

The long-term strength of the Ni-2ThO$_2$ alloy due to the effect of dispersion hardening of ThO$_2$ appears at temperatures above 1 000 ℃. The alloy at high temperatures has 2—4 times higher high-temperature strength than the alloys based on nickel and cobalt and is inferior only to refractory metals. The heat resistance of such alloy is 3—5 times higher than that of nickel, and its heat and electrical conductivity correspond to pure Ni.

The long-term strength of the Ni-2ThO$_2$ alloy is almost independent from time. The time to destruction at temperature of 760 ℃ is 843 hours, and at temperature of 1 200 ℃ is 706 hours, while R_{100}^{900} = 100—110 MPa. The addition of chromium to such alloys improves the performance and dispersion hardening efficiency[115].

Based on chemical and metallurgical methods for producing powders using deformation-heat treatment, VIAM (Russia) developed dispersion-strengthened alloys with nickel (VDU1 and VDU2) and nickel-chromium matrix (VDU3) using non-toxic HfO$_2$ (in VDU2 and VDU3 alloys). Chemical and metallurgical methods for producing powders of dispersion-strengthened nickel-based alloys are based on carbonate coprecipitation of poorly soluble nickel compounds and a hardening phase from nitric acid solutions, followed by thermal decomposition of precipitates and selective hydrogen reduction. However, it should be noted that this technology is not applicable to alloys containing highly active alloying elements (Al, Ti, Nb, etc.).

The maximum high-temperature strength is reached for rods from VDU1 alloy: $R_{100}^{1\,100}$ = 100 MPa; $R_{100}^{1\,200}$ = 80 MPa; $R_{100}^{1\,300}$ = 50 MPa. The sheets have lower high-temperature strength, for example for VDU2: $R_{100}^{1\,100}$ = 50 MPa; $R_{100}^{1\,200}$ = 30 MPa, which is nevertheless several times higher than that of traditional sheet materials. A significant advantage of VDU1 and VDU2 alloys is their high thermal conductivity ($\lambda > 70$ W/(m·K)), which is associated with the absence of alloying additives in the matrix. A common disadvantage of alloys of the VDU series is their low strength, at room and medium temperatures (up to 800 ℃), due to limited alloying possibilities. This disadvantage has been successfully overcome in dispersion-hardened alloys obtained using mechanical alloying.

Dispersion-hardened superalloys can be used up to temperatures of 1 100 ℃. Such materials are used to manufacture fixed guide vanes located at the entrance of each turbine circuit. The sheet materials from dispersion-hardened superalloys are promising for using as plating in the combustion chamber. The presence of dispersed oxides ensures high strength of these materials due to the dislocation pinning mechanism (Orowan strengthening) at higher temperatures than the effective hardening effect area of the γ'-phase precipitates. In addition, such alloys are distinguished by high corrosion resistance in aggressive media due to the formation of a dense oxide layer on their surface, which is well adhered to the substrate[112].

Dispersion-strengthened superalloys powders are produced by *mechanical alloying*. This solid-phase process makes it possible to obtain alloy powder from almost any composition from a powder mixture of pure metals, oxides and master alloys. Grasping is a diffusion-free process that involves the formation of a physical contact (that is, a convergence of interacting atoms, during which valence interatomic interaction can occur) and activation of contact surfaces that implements interatomic interaction on surfaces.

Mechanical alloying is carried out in special ball mills, more often in attritors. The process can be represented as repeated acts of welding the powder components onto the surface of the balls, redistributing the components under the influence of high-speed microplastic deformation, cleaving the extremely cold-worked composite particles from the surface of the balls and their spheroidization. As a result, each particle is like a micro-ingot of the alloy, and, unlike the granules obtained by spraying, in the mechanically alloyed powders submicron oxide reinforcements are evenly distributed in volume.

The technological process of mechanical alloying is used to obtain a powder of alloys with the structure of austenitic solid solution (MA-754), ferrite (MA-956) or γ'-hardened austenite (MA-6000) (see Table 5.4). MA-6000 alloy is distinguished by good strength at medium temperatures, where the strength of MA-754 and MA-956 alloys is insufficient[112].

TABLE 5.4 Chemical compositions of alloys hardened by dispersed oxides

Alloys	Concentration of chemical elements/wt. %											
	Fe	Ni	Cr	Mo	W	Al	Ti	Ta	B	Zr	C	Y_2O_3
MA-754	—	Bal.	20.0	—	—	0.3	0.5	—	—	—	—	0.6
MA-956	Bal.	—	20.0	—	—	4.5	0.4	—	—	—	—	0.5
MA-6000	—	Bal.	15.0	2.0	4.0	4.5	2.5	2.0	0.01	0.15	0.05	1.1

The main way to consolidate dispersion-strengthened powders of superalloys is hot extrusion. Due to chemically passive nature of such powders, they do not need to be protected during processing. The powders are compacted with uniaxial pressing or CIP. The pressed billet or powder is placed in a container, heated and extruded with deformation degrees sufficient to produce a completely dense material. At the same time, the final product (due to the uniform distribution of oxides) has a fine-grained ($\sim$1 μm) structure. The resulting billet is subjected (if necessary) to thermal or thermomechanical processing. In any case, the final operation is recrystallization annealing usually at temperature of 1 260 ℃, as a result of which stable (due to the fixing of boundaries with inert oxides) and elongated grains grow. Often each grain can have a length of up to 10 μm. Such grain size and their preferential orientation (texture) provide exceptional high temperature stability and strength of such materials by minimizing the length of grain boundaries in the transverse direction.

Alloys hardened with dispersed oxides undergo thermomechanical treatment, which is typical for all superalloys (at temperature of 815—1 100 ℃), and in consolidated state they have a fine-grained structure and have superplastic properties. Thermomechanical processing is carried out with the purpose of hot deformation of the material to the critical value necessary for its recrystallization. Deformation is usually carried out on a hot tooling, which allows to use the material more economical and provide necessary control of the deformation process itself. After deformation for the growth of grains of the desired shape and size, it is recrystallized. To enhance the appropriate directionality of the structure, recrystallization is often carried out under conditions of temperature gradient. After that, to obtain the required size distribution of γ'-precipitates and carbides, alloys of the MA-6000 type undergo the same heat treatment as standard superalloys, strengthened by precipitates of the γ'-phase[112].

The presence of a pronounced crystallographic texture causes the anisotropy of the mechanical and physical properties of dispersion-strengthened superalloys. In cases where good thermal fatigue resistance of the material is required (for example, for gas turbine guide vanes), in the MA-956 alloy it is possible to create a strong texture when deformed in parallel to the $<100>$ crystallographic direction. This texture leads to a low value of the elastic modulus in the longitudinal direction, which improves the resistance of thermal fatigue due to the reduction of stresses at a given level of thermal deformation. Anisotropy of mechanical properties is also noticeable during short-term tensile tests; however, it is much more pronounced when studying creep and long-term strength, that is, time-dependent properties. Proper use of organized texture and knowledge of tensions that influence the structure allows improving the performance of specific parts. The mechanical properties of disperse-strengthened superalloys in the longitudinal and transverse direction are given in Table 5.5.

TABLE 5.5 Mechanical properties of dispersion-strengthened superalloys

Alloy		Tensile test at $T=1\,100$ °C				$R^{1\,100}_{1\,000}/$ MPa	$R^{800}_{100}/$ MPa
		$R_{100}/$MPa	$R_{p0.2}/$MPa	$A/\%$	$Z/\%$		
MA-754	in longitudinal direction	148.2	134.4	12.5	24.0	93.8	—
	in transversal direction along the long axis	131.0	120.6	3.5	1.5	24.1	—
MA-956	in longitudinal direction	91.0	84.8	3.5	—	—	66.9
	in transversal direction	89.6	82.7	4.0	—	—	62.7
MA-6000	in longitudinal direction	222.0	191.7	9.0	31.0	110.3	—
	in transversal direction	177.2	170.3	2.0	1.0	—	—

MA-956 alloy was developed as a high-temperature material to be used in the form of sheets. The advantage of this alloy is its excellent oxidation resistance. MA-6000 alloy was developed as a material that combines high-temperature strength at intermediate temperatures, characterized as alloys hardened by the γ'-phase precipitates[112].

The basis of the mechanically doped dispersion-strengthened alloy VPM-1 is the Ni-Cr-Al-Co-W-Nb system with the addition of 0.5% Y_2O_3 (no elements and micro additives were introduced into the alloy to decrease the temperature of solvus and solidus-Ti, Mo, Zr, B). To ensure scaling resistance, the chromium content in the alloy was limited to 5%. As a result, at fraction of about 55% of the γ'-phase, the alloy has solvus and solidus temperatures of 1 230 °C and 1 400 °C, respectively. The semi-finished products deformed by extrusion and hot rolling after zone annealing hardening with aging are characterized by structure with strongly elongated coarse grains and fairly fine gamma-γ precipitates. The mechanical properties of the VPM-1 alloy rods are presented in Table 5.6[116].

TABLE 5.6 Mechanical properties of rods from VPM-1 alloy

$T/℃$	$R_{p0.2}/MPa$	R_m/MPa	$A/\%$	$Z/\%$	R_{100}/MPa
20	1 100—1 200	1 400—1 500	9—14	8—12	—
100	300—330	330—360	1—2	1—2	150—170
1 000	180—200	190—230	3—6	2—4	100—120
1 200	100—120	110—130	3—6	2—4	50—60

The VPM-1 alloy has unique high heat resistance: an increase in mass per 100 hours during oxidation in air is ~4 g/m² at temperature of 1 100 ℃ and 6.5 g/m² at temperature of 1 200 ℃. Due to the high creep resistance and oxidation, VPM-1 alloy can be used for high-temperature fasteners, uncooled rotor blades and other parts with operating temperature of 1 250—1 300 ℃[116]. Studies of the dispersion-strengthened multicomponent alloy with the composition of Ni-8Cr-6.5Al-6W-3Ta-3Re-1.5Mo-5Co-1Ti-0.15Zr-0.01B-0.05C, hardened with 1.1% Y_2O_3, which is introduced by mechanical alloying, are being performed. The additions of Ta and Re also significantly improve the high-temperature strength of modern nickel alloys.

In addition to nickel alloys, dispersion-strengthened high-temperature strengthened steels were developed. Chromo-aluminum ferritic steel Fe-23 Cr-5Al-1Ti, hardened with dispersed oxides by mechanical alloying, is used as material operating at temperatures over 900 ℃, which is impossible without dispersive hardening due to the strong softening of the steels.

The dispersion-hardened steel VPM-2 (Fe-Cr-Al matrix with Ti and Nb, addition of 0.5% Y_2O_3) has a melting point $T_{mel}=1\,480\,℃$, $\rho=7\,260\,kg/m^3$ and can be used at temperatures of 1 300—1 350 ℃. Due to the good processability of this steel, it is possible to produce sheets, foil, rods, wire, roll rings and other semi-finished products. A feature of steel is the absence of degradation of properties during prolonged heating. Tested after heating at temperature of 1 200 ℃ for 100 h, the samples did not show a drop in tensile strength and relative to elongation. The mechanical properties of steel VPM-2 are presented in Table 5.7[116].

TABLE 5.7 Mechanical properties of semi-finished products from steel VPM-2

$T/℃$	$R_{p0.2}/MPa$	R_m/MPa	$A/\%$	R_{100}/MPa
20	500—600	700—800	8—13	—
100	80—90	95—110	4—7	70
1 000	70—80	90—100	3—5	50
1 200	50—70	80—90	3—5	35

This steel, like all dispersed-hardened materials, is characterized by a slow rate of softening during long-term strength tests, therefore, it is advisable to use it in durability products. High heat resistance (an increase in mass per 100 hours of heating in air environment does not exceed 10 g/m^2 and 20 g/m^2 at temperatures of 1 100 ℃ and 1 200 ℃, respectively) eliminates the problem of protection against oxidation. It can be used in constructions of combustion chambers, nozzle assemblies and nozzle apparatus of aircraft engines, as well as the material for the front edges of the wing and the tail in hypersonic aircraft.

5.4 Refractory Metal Compounds (Intermetallic Compounds)

Intermetallic compound is a chemical combination of metals with each other. Unlike conventional chemical compounds, intermetallic compounds often do not obey stoichiometry—the laws of constant composition and multiple proportions.

Intermetallic compounds and alloys are considered as promising high-temperature strengthened materials for the most critical purposes, in particular in aviation and space technology. The basis for such assessment is the following characteristics of intermetallic compounds:

① Intermetallic compounds retain high strength up to fairly high temperatures; moreover, the strength properties of some intermetallic compounds increase with increasing temperature.

② The elastic modulus of intermetallic compounds less actively decreases with increasing temperature than the similar properties of alloys with crystalline structure.

③ The coefficient of self-diffusion in intermetallic compounds is in several orders of magnitude lower than that of alloys at comparable temperatures; on this basis, one can expect a significantly lower creep rate in intermetallic compounds.

④ Intermetallic compounds containing light elements (Al, Be, Ti, Si) as basic elements have low density, which provides high specific strength properties, especially at elevated temperatures, which is crucial for aviation and rocket technology.

In addition, intermetallic compounds have good anticorrosive properties in various environments, which provide prerequisite to consider them as promising material to replace superalloys.

For aerospace engineering, lightweight high-temperature strengthened alloys based on intermetallic compounds of titanium with aluminum, known as titanium aluminides, are of particular interest. In the Ti-Al binary system, there are three chemical compounds—Ti_3Al (α_2-phase), TiAl (γ-phase) and $TiAl_3$.

Up to temperatures of 700—800 ℃, the intermetallic compounds α_2-Ti_3Al (15.8Al) and γ-TiAl (36% Al) remain high strength, elastic modulus and heat resistance. Due to their low density, the specific properties at high temperature exceed those of titanium and nickel alloys (see Table 5.8). The main disadvantage is low plasticity at room temperature.

Titanium aluminides are about 3 times cheaper than nickel alloys that work for a long time at temperatures of up to 700 ℃, and they are 5 times cheaper than the corresponding nickel alloys per unit specific strength[113].

Melting and casting of intermetallic compounds is associated with a number of technological difficulties—it is necessary to strictly observe the chemical composition, in order to avoid cracking of the ingot, and additional measures are necessary in the cooling process, which negatively affects the formation of the optimum, in terms of properties, structure. The formation of a coarse-crystalline structure worsens the already poor plastic properties of intermetallic compounds. High resistance to deformation makes it difficult to manufacture forgings and, especially, sheets and foils. Machining requires the use of carbide tools and meets almost the same cutting conditions as for nickel alloys.

In this connection, powder metallurgy methods are promising. Powders are obtained by centrifugal spraying of melts or by mechanical alloying. After HIP, mechanically alloyed powders have nanostructure and remain it up to temperature of 600 ℃. The existence and high thermal stability of the nanostructure significantly improve their properties[113]. Particular attention is currently being paid to TiAl-based intermetallic compounds additionally alloyed with Nb and other elements (Cr, Ni, B, C) to improve their performance[92]. Table 5.8 gives a comparison of the high temperature mechanical properties between the TiAl-based intermetallic compounds and the titanium and nickel alloys.

TABLE 5.8 Comparative properties of alloys with high-temperature strength[117]

Alloys	ρ/ (kg·m^{-3})	E/GPa	$T_{work(max)}$/℃		A/%	
			By creep	By oxidation	At 20 ℃	Operating temperature
Titanium TC4	4 500	96—120	540	550	10	high
Ti_3Al basis	4 150—4 700	110—145	815	650	2—5	5—8
TiAl basis	3 760	176	1 040	1 040	1—2	7—12
Nickel	8 300	206	1 095	1 095	3—5	10—20

As it is known, the grinding of the microstructure contributes to the plasticization of brittle metals. In this regard, mechanical alloying can be an effective plasticization technology for intermetallic alloys, when the alloying process takes place simultaneously with the grinding of a mixture of powders during mechanical grinding—up to the formation of nanoscale structure. Such approach was used to develop an intermetallic alloy of the Nb-Al system, intended for use in aerospace engineering. After compaction, the Nb-Al intermetallic samples show plasticity at the level of standard nickel alloys but have an operating temperature above the melting point of nickel alloys with high-temperature strength ($R_m^{1\,600}$ = 185—200 MPa).

Powder metallurgy intermetallic alloys based on Ti, Fe, Ni and Nb aluminides belong to the category of materials of a new generation which are interesting to be used in aerospace engineering.

The Ni-Al-Cr-W-Mo-Ti-Ta-C-(RE) alloys developed based on the Ni_3Al alloying system did not find any use in the cast state due to its increased brittleness and was used as material for heat-resistant and wear resistant coatings for many years. The alloy produced by powder technology has satisfactory cold plasticity and high strength properties at elevated temperatures ($R_{100}^{1\,250}$ = 14 MPa). This alloy is characterized by the fact that it can be welded by argon-arc and electron-beam welding both with itself and with austenitic stainless steel 0.12C-18Cr-10Ni-1Ti.

Promising heat-resistant materials for aerospace engineering are *beryllides*—high-temperature intermetallic compounds of beryllium with other metals. RE beryllides have T_{mel} = 1 720—1 920 ℃. The beryllides with transition metals are distinguished by high strength, thermal conductivity and resistance to oxidation at high temperatures (1 200—1 500 ℃). Beryllide Ta_2Be_{17}, which has good high-temperature properties, including antifriction property, seems to be one of the most suitable candidates to meet the needs of aerospace engineering[111].

5.5 Special Ceramic Materials with High Temperature Strength

Special Ceramic Materials are polycrystalline materials obtained by pressing and sintering (or hot pressing) powders of non-metallic carbides and nitrides, as well as their mixtures, sometimes with addition of small amounts of refractory oxides. The starting materials are characterized by high chemical inertness in reducing environments, and SiC and Si_3N_4 possess outstanding heat resistance at high temperatures due to the formation of a dense stable oxide film on their surface during oxidation. Another feature of these materials based on non-metallic refractory compounds is their low density ($\rho \leqslant 3\,000$—$3\,500$ kg/m^3). Another advantage of ceramic materials is their low price and wide distribution of raw materials (Si, C and N) in nature[112].

The disadvantages of ceramics include their fragility, sensitivity to thermal shock, and lower thermal conductivity than heat-resistant metallic materials. In this regard, such materials work very poorly under the influence of tensile stresses. The mechanical properties of ceramics are highly dependent on the method of manufacturing of the product, the impurity of the material and the cleanliness of the surface. Nevertheless, the viscosity and resistance to thermal shocks, as well as the ability to form protective oxide layers are sufficient for the Si_3N_4 and SiC compounds to become candidates for use in hot turbine assemblies. The reduction of the weight of products when using ceramics instead of metal alloys should be noted.

Massive ceramic products in the classic version are made from powders. With equal bulk density, the hot-pressed material is stronger than that which is sintered in the traditional way or obtained as a result of thermal synthesis. However, hot pressing is not suitable for manufacturing of products with complex shape and does not provide high performance.

Features of the process of producing ceramic materials have higher impact on the final properties of products than in the production of metallic materials. The scatter in the properties of silicon nitride products of Si_3N_4 with the same raw materials is determined by the need to use oxides additives that promote the sintering of high-purity Si_3N_4 powders. Additives concentrated at grain boundaries form liquid silicates as a result of combining with SiO_2, which is usually present in Si_3N_4 powders. Compaction occurs as a result of dissolution and separation of the second phase in the liquid silicate. Thus, the material properties are determined by the chemical composition and volume fraction of silicate remaining at the grain boundary, and its applicability is limited by the temperature at which the boundary material loses its strength. For MgO additives in Si_3N_4, this temperature is close to $1\,300$ °C. The oxides of rare-earth elements, Y_2O_3 in particular, are shown to effective additives for promoting sintering. The Si_3N_4-Y_2O_3 system has demonstrated excellent properties, especially at temperatures above $1\,200$ °C. However, in some compositions phase instability is detected in the temperature range of 700—$1\,100$ °C[112].

Hot Isostatic Pressing (HIP) are utilized to provide the system Si_3N_4-Y_2O_3 with high density. A combination of cold isostatic pressing followed by HIP without envelope is promising for SiC and Si_3N_4[112].

With sufficient chemical purity and processing temperature, the density of Si_3N_4, which is close to the theoretical density, can be obtained without "sintering" additives. Sintering at temperatures of 1 600 ℃ and 1 800 ℃ and pressures of 1—5 GPa provides a relative density of 0.88 to the theoretical one. As a result, the hardness increases by ~50% compared with the hardness of Si_3N_4-4Y_2O_3[112].

A sintering process with nitriding without pressure has been developed for the manufacture of Si_3N_4 ceramics. For this, a ceramic blank is molded, and then it is nitrated. Free silicon combines with nitrogen and turns into a Si_3N_4 compound at temperatures of 1 300—1 400 ℃. The final sintering is carried out at temperature of 1 800 ℃. Linear shrinkage is 8%—9%, whereas in normal sintering it is 15%—20%[112].

Polycrystalline materials based on Si_3N_4 have a low Coefficient of Linear Thermal Expansion (CLTE). $\alpha = 2.8 \times 10^{-6}$ K^{-1} (for the sintered material) and $\alpha = 3.2 \times 10^{-6}$ K^{-1} (for the hot-pressed material). When heated to 1 000 ℃, the fall in the elastic modulus E is not observed. The flexural strength of reactive polycrystalline Si_3N_4 is almost linearly related to its density (porosity), that is $R_{m,ben} = 83$ MPa with $\rho = 1\ 800$ kg/m^3 and $R_{m,ben} = 216$ MPa with $\rho = 2\ 500$ kg/m^3. Reactive sintered Si_3N_4 has $R_m = 50 - 60$ MPa with $\rho = 2\ 240$ kg/m^3. As the temperature rises to 1 100—1 300 ℃, the strength of ceramics based on Si_3N_4 does not change, which is very important when using it as a material that is strong at high-temperature[118]. In spite of the low thermal conductivity of Si_3N_4, it is very heat-resistant and does not fail during rapid heating and cooling (at a speed of ~1 000 ℃/s).

Materials based on Si_3N_4 are promising for the manufacture of turbine blades, combustion chambers, flame tubes and other Gas Turbine Engine (GTE) parts operating at temperatures above 1 370 ℃. High-temperature strength and heat resistance of dense polycrystalline Si_3N_4 meet the requirements for materials of such products. A relatively weak point of the material is the impact strength, which can be significantly improved by reinforcing nitride ceramics with fibers of refractory metals (for example, Ta). A more serious problem is the optimization of manufacturing technology for products of complex shape. The use of Si_3N_4 submicron powders by sintering to replace hot pressing is promising[118].

Silicon carbide SiC in α-(hexagonal) and β-(cubic) crystal structure is obtained by chemical reaction, sintering or hot pressing. Hot pressing provides the hardest and most viscous SiC product. Ceramic material SiC has higher heat resistance than Si_3N_4. SiC is considered as high-temperature structural material, suitable in particular for gas turbine blades with a working temperature above 1 400 ℃. As in case with Si_3N_4, sintering of submicron SiC powders with additives B and C is promising, as well as the combination of hot pressing with controlled sintering, in which the β-SiC to α-SiC transition is suppressed[112].

Chapter 6　Composite Materials

6.1　General Concepts About Composite Materials

A *Composite Material* (*CM*) is a combination of two or more chemically dissimilar materials with a clear boundary between them and it has properties that none of its individual components can possess. Composite materials are heterogeneous in nature.

There are several classification schemes for composite materials. The most commonly used classifications are based:

① on the structure of the material (poly-matrixes and poly-reinforcements);

② on the form of the filler (zero-dimensional, one-dimensional and two-dimensional);

③ on the reinforcement dimensions (uniaxial, biaxial and triaxial).

According to the structure of the composite material, the poly-matrix and poly-reinforced materials are distinguished. Fig. 6.1 shows the schemes for constructing poly-matrix and poly-reinforced composite materials with uniaxial strengthening by one-dimensional filler (fiber).

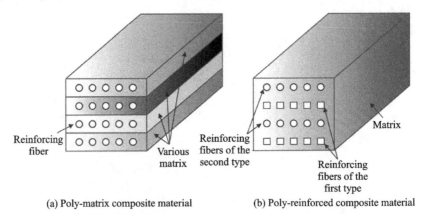

(a) Poly-matrix composite material　　　(b) Poly-reinforced composite material

FIGURE 6.1　Composite materials

Based on the shape of the reinforcing filler, similar to the classification of nanostructures (as was shown earlier in Chapter 3), D_0, D_1 and D_2 are distinguished, that is, zero-, one-, two-dimensional fillers can be used (see Fig. 6.2).

Based on the reinforcement scheme, the composites are divided into three

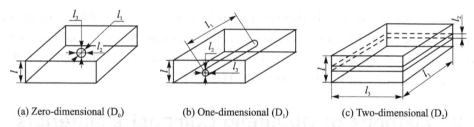

(a) Zero-dimensional (D_0)　　(b) One-dimensional (D_1)　　(c) Two-dimensional (D_2)

FIGURE 6.2　The fillers added based on the form

groups ($kDlmn$), where k is the dimension of reinforcement, l, m, n are the dimension of reinforcing filler (see Fig. 6.3).

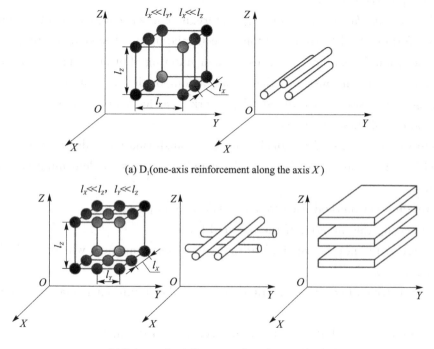

(a) D_1(one-axis reinforcement along the axis X)

(b) D_2(two-axis reinforcement along the axes X and Y)

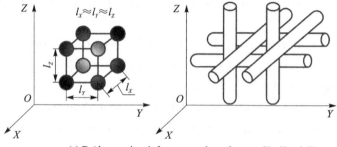

(c) D_3(three-axis reinforcement along the axes X, Y and Z)

FIGURE 6.3　Reinforcement schemes for composites

The composite materials themselves can be two-dimensional (2D) (sheets) or three-dimensional (3D) (massive products). In accordance with the classification of 2D composites according to the reinforcement scheme there can be 2D1 and 2D2, and 3D composites can be 3D1, 3D2 and 3D3.

6.2 Methods of Obtaining Composite Materials

All existing methods of obtaining composite materials can be divided into two classes—*indirect* and *direct*.

Indirect methods: hot pressing, impregnation with liquid metal (including vacuum forming). Electrodeposition (electroplating), plasma deposition, vapor deposition, ion deposition, cold pressing and sintering, hot pressing, extrusion and impregnation, joining methods that create a composition by pulsed energy supply (forging, explosion, electric shock, spark plasma sintering, etc.), various combinations of these methods.

Direct methods: directional eutectic transformation, directional eutectic crystallization, creation of fibrous and layered structures by deforming two-phase systems.

Any indirect method used to obtain CM should provide:

① getting the material of the required form;

② integrity (not destruction) of the reinforcing component when introducing it into the matrix;

③ no technological interaction between the reinforcing material and the matrix;

④ uniform distribution of the reinforcements in the matrix;

⑤ the possibility to introduce a significant amount (share) of the reinforcing component, which leads to significant strengthening;

⑥ achieving the required strength at the interface of the matrix—reinforcing phases.

Based on these requirements, and in order to reduce technological (production) costs, solid-phase methods are usually used for simple parts: sintering, hot pressing, free forging, rolling, explosion welding.

To obtain parts of a complex spatial form, liquid-phase methods are used: impregnation, drawing through the melt, casting, as well as coating deposition methods (see above).

6.3 Metallic Structural Composite Materials

Metallic composite materials, in which the matrix is aluminum, magnesium and titanium alloys, are one important classification of composite materials that the aerospace industry today are interested in. The reinforcing phases can be discrete fillers (thread crystals "whiskers"—Al_2O_3, SiC, carbon—including fullerenes), nanopowders or continuous fibers (boron, carbon or steel wire).

The most developed compositions in aircraft construction are steel – aluminum, SiC – aluminum and boron – aluminum composite materials.

As an example, Figure 6.4 shows a technological scheme of obtaining a boron – aluminum composite from boron fiber. Boron fiber is obtained from volatile boron compounds by decomposition on a tungsten wire, which is then wound with a certain step on the drum in a single layer and fastened together with aluminum, which is sprayed with a plasma method.

The resulting monolayer is cut across the boron fibers. This is followed by installing and hot pressing with subsequent finishing operations to make the finished product. A similar technology is used in manufacturing other composites with fibrous fillers, e.g., the steel – aluminum composite, where thin high-strength steel wire is used as reinforcement.

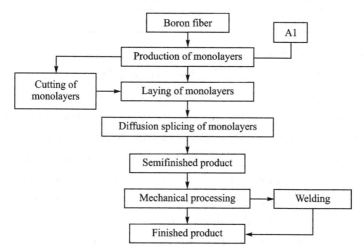

FIGURE 6.4 Technological scheme of obtaining the composite Al – B

Composite materials hardened with discrete reinforcers (nanopowders, whiskers, nanotubes, fullerenes and micropowders) have the ability for standard types of processing—hot (sometimes cold) deformation, cutting, etc., and also have isotropy of physical and mechanical properties. By changing the composition of the material, introducing certain reinforcements, in particular in aluminum alloys, it is possible to purposefully change not only the strength properties, but also the physical ones—the elastic modulus, thermal expansion coefficient, thermal conductivity, etc. Composites of aluminum alloy (2XXX system) reinforced with SiC have a thermal expansion coefficient ($= 13 \times 10^{-6}$ K^{-1}) at the level of structural steels, which is very important for precise instrument making and relevant for internal-combustion engine pistons. The addition of reinforcements such as boron, carbon, and SiC significantly increases the elastic modulus (up to $E = 250$ GPa against $E = 70$ GPa for aluminum), which is extremely important for parts working for buckling (compression), for example, support for optical space antenna mirrors. Properties of composite materials of the 2XXX – SiC system are presented in Table 6.1.

TABLE 6.1 Physical and mechanical properties of composite materials based on commercial alloys of the Al-Cu-Mg system

Content SiC/ vol. %	$\rho/(\text{kg} \cdot \text{m}^{-3})$	E/GPa	R_m/MPa	δ/%	$\alpha \times 10^6$/K ($T = 20$—100 K)
15	2 840	90—100	500—600	5—7	19
20	2 860	100—110	600—650	3—4	17
30	2 900	110—120	520—670	1	14
40	2 940	130—140	620—670	1	12

Among structural composite materials, we should mention a relatively new material based on the titanium matrix (for example, TC4) reinforced with the TiB$_2$ system. Such material is obtained by direct method of eutectic crystallization or by sintering a powder mixture of the appropriate chemical composition. The creation of this composite is based on the property of titanium to form borides, but not to dissolve boron in the matrix. The second premise is the presence of eutectic in the titanium-boron system. The specified composite can be obtained both by melting and by powder metallurgy, with the second option being preferable due to the increased uniformity of the distribution of the components and smaller boride precipitates. Introduction to the standard TC4 alloy with 1.55 wt. % B allows to increase its strength from $R_m \leqslant 950$ MPa to $R_m \leqslant 1\,440$ MPa, as well as the improved wear resistance. The microstructure of the TC4 composite with addition of 1.55 wt. % B is shown in Fig. 6.5.

Currently, a number of composite materials on titanium matrix have been developed, but they have not been found wide application yet. The main difficulty

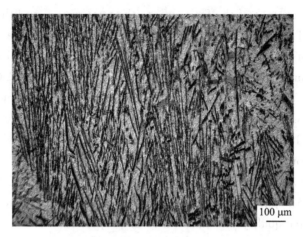

FIGURE 6.5 The microstructure of the TC4 composite with addition of 1.55 wt.% B produced by powder metallurgy

in obtaining composite materials based on titanium is due to the fact that titanium at elevated temperatures reacts with almost all substances and thereby the matrix is embrittled. Magnesium-based composite materials do not spread due to the low corrosion properties of the matrix.

It should be noted that in the aerospace industry there exist in-use composite materials that are not produced by powder metallurgy methods—these are laminated materials obtained by joint hot deformation. These include aluminum-titanium cladding sheets used as aircraft linings in areas with increased erosion wear (in the landing gear area), and aluminum-copper tires used as electrical wires for high-frequency electrical systems for the aircraft. Laminated aluminum – glass fabric composites (Allor, Gler), which have a significant advantage over aluminum alloys in terms of crack resistance, are also used. They are used for producing the outer skin of the aircraft in areas with high vibration loads (in the area of engines).

One class of structural and special composite materials that have begun to be used in aerospace engineering is porous materials, in particular, the most common foam aluminum. In these composites, the role of the matrix is played by the metal (alloy), and the filler gas. For the first time a method of producing foam aluminum was developed in Germany in the 90s of the 20th century. The essence of the method is foaming of powder briquette made from metal powder (for example, aluminum) and foam agent. Foaming occurs when the briquette is heated to temperatures close to the melting point of the metal component. At this temperature, the foam agent dissociates, and the released gas foams the preform.

The time for foaming the workpiece is from several seconds to several minutes. The relative density of the foam is controlled by the following parameters—the content of the foam agent, the temperature and the heating rate of the workpiece. The choice of the foam agent is determined by the proximity of the temperature of its dissociation to the melting point of the metal component. To obtain aluminum foam, hydrides (TiH_2, ZrH_2) and/or $CaCO_3$ are used. Currently, there are two main methods of producing foam products—obtaining a semi-finished product by foaming a briquette in a metallic form of predetermined dimensions, and obtaining foamed granules followed by the manufacture of products therefrom.

The second method is more preferable—it is a new technology, called *Advanced Pore Morphology (APM) technology*, which allows to make sandwich panels with higher quality and uniformity of properties[119]. When manufacturing small-sized foamed granules, it is quite easy to control the process of pore formation and prevent the coalescence of pores, which is inevitable when a large amount of material is foamed.

Then, the granules are glued together using a sandwich panel of the desired design (see Fig. 6.6 (b)). This method not only improves the quality of the products, but also makes them cheaper.

 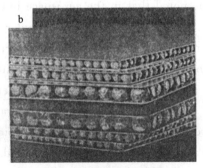

(a) Aluminum foam in a metal form (clad)   (b) Construction consisting of sheets and APM–foam components

FIGURE 6.6 Sandwich Panels

The combination of low density and unique physic and mechanical properties of foams determine the prospects of their use as multifunctional structures in aerospace engineering. Foam aluminum looks especially attractive if we compare its weight characteristics with other materials with the same structural rigidity. So, a steel sheet construction with a thickness of 0.8 mm will be 6.24 times heavier than a 10 mm-thick foam aluminum construction with a density of 100 kg/m^3 and 4.5 times heavier than foam aluminum construction with a density of 700 kg/m^3 [120].

The use of foam aluminum in various energy absorbers is due to the fact that the deformation diagram during its compression has a lengthy horizontal section

(55% to 70% before deformation), that is, aluminum foam is destroyed almost at constant load. To increase the length of the horizontal section and smooth it, the specialists use foam aluminum granules as a filler for buffer devices. An additional advantage of granules compared to "monolithic" foam aluminum when manufacturing buffer devices is the convenience of filling the shells of complex geometric shapes.

Foam aluminum can provide protection against sound vibrations (sound insulation) through a combination of two processes—reflection and absorption, depending on the type of porosity (closed or open). Foamed aluminum with closed porosity largely reflects sound vibrations, and with open porosity absorbs them. The effectiveness of noise protection panels can also be significantly improved by the use of bulk filler in the form of foamed granules in the design.

Due to its good ability to absorb electromagnetic energy, foam aluminum can be used in structures that attenuate electromagnetic radiation. Up to frequency of 10 MHz, the efficiency of foam aluminum screen is similar to that of a silicon steel screen, at higher frequencies the efficiency of foam aluminum is higher.

When using aluminum foam as an insulating material, its advantage is higher strength properties not only at normal but also at higher temperatures. It meets the requirements of aviation standards for flammability and toxicity (there is no release of toxic gases) in the event of a fire. At the same time, it can also serve as a construction material, for example, for manufacturing partitions between the cockpit and the passenger cabin of the aircraft.

Foam aluminum has a low hygroscopicity (1%—3%), so its properties depend little on this property, causing frost and cracking resistance when the temperature drops. In this way, it compares favorably with composite materials on an organic matrix.

6.4 Heat-resistant Composite Materials

In heat-resistant composite materials the matrix material must have significant resistance to oxidation, high strength at elevated temperatures, satisfactory plasticity at room temperature and should not react at operating temperature with reinforcing fillers.

Metallic heat-resistant composites is a class of composite materials in which the matrix is metals or their alloys (Ti, Al, Cu, Ni, Cr, etc.), and the filler is discrete or continuous fibers of metals, non-metals and various compounds, single crystals, particles, nanoparticles, nanofibers, whiskers.

At present, a number of composite materials on a titanium matrix reinforced with SiC, Nb and B fibers, which are very promising for use in jet engines at operating temperatures up to 600 ℃, have been studied. An example is the composite material SCS-6, obtained on the basis of the IMI834 pseudo-α alloy (Ti-5.8Al-4Sn-3.5Zr-0.7Nb-0.5Mo-0.35Si) reinforced with SiC fibers. To prevent chemical interaction at operating temperatures, the fibers of the hardener are protected by a barrier coating.

For parts of gas turbine engines operating at temperature of 350—400 ℃, composite materials based on aluminum and magnesium alloys with B and SiC fibers with high strength ($R_m = 1\,200$—$1\,400$ MPa) and the modulus of normal elasticity ($E = 230$ GPa) are promising. According to these properties, they exceed (at the indicated temperatures) aluminum and titanium alloys by 2—3 times in strength and rigidity[121].

For operation at temperatures of 1 100—1 300 ℃ competitive materials for nickel superalloys are composite materials based on heat-resistant chrome matrix reinforced with refractory oxide or tungsten fibers. Such composites have a heat-resistance of $R_{100}^{1\,200} = 200$ MPa and $R_{100}^{1\,300} = 100$ MPa[122].

Promising for aircraft engine building are Ni-Al intermetallic composites, reinforced with fibers, including nanoscale oxides, which have heat-resistance of $R_{100}^{1\,200} = 80$—100 MPa and oxidation resistance. Compounds of the Nb_3Al-W system are recommended for operation at temperatures above 1 300 ℃[122].

Further improvement in performance of metallic composite materials will be possible when obtaining reinforcing components with higher heat-resistance, the search for which is carried out in the direction of nanoscale fibers, in particular boron, carbon, carbide and oxide.

Ceramic composite materials (*ceramic composites*) is a class of composite materials, whose matrix is ceramics (O_x, C_x, N_x, B_x, Si_x) compounds and the reinforcing hardener is fibers or particles of metals, compounds and alloys[123].

Ceramic composites are considered as super heat-resistant materials. Ceramic materials based on oxide matrix are efficient in such operating conditions, where even the most heat-resistant metals and alloys tend to oxidize and creep.

Aluminum oxide Al_2O_3 is the most common heat-resistant ceramics, which has excellent chemical resistance and heat-resistance properties, but is brittle for use as structural heat-resistant material in aerospace products. To reduce the brittleness of Al_2O_3, aluminum oxide-based composite materials have been developed and continue to be developed, containing various hardeners—refractory metals and alloys, intermetallic compounds, ceramic fibers and powder materials. Composites reinforced with metal obtain plasticity that is not typical for ceramics, which increases its performance. Intermetallic compounds—TiAl, TiAl, Ni_3Al, Fe_3Al, NbAl and others are also an effective hardener for ceramics. Ceramic-ceramic composites with the highest hardness are also used, where SiC, TiC, ZrO_2, and others are used as hardeners.

Ceramics with eutectic compositions Al_2O_3-ZrO_2, Al_2O_3-ZrO_2-Y_2O_3, which has enhanced performance properties due to the inhibition of cracks at the interface between the components of the eutectic, has received widespread use. Composite materials of this type retain their strength properties up to temperatures of 1 200— 1 300 ℃.

The most promising heat-resistant composite materials are *carbon-carbon and carbon-ceramic composite materials*, which can operate at temperatures up to 2 000 ℃, however, not in an oxidizing environment. Such materials are superior to structural ceramics in heat-resistance by 5—10 times. Currently a lot of work is being carried out all over the world in this direction[124-126].

Ultra-high-temperature ceramics is a composite with the matrix ZrB_2 and/or HfB_2, dispersed-strengthened by inclusions of SiC particles and other refractory compounds (carbides, silicides, nitrides) and the structure protecting its surface from oxidation. The most promising is the ZrB_2-SiC system, which combines high oxidation resistance and erosion resistance at high temperatures with mechanical strength. To increase the heat resistance of the specified composite ceramics can be added to the initial mixture of $ZrSi_2$ powders. Such ZrB_2-SiC-$ZrSi_2$ ceramics has a 2.7 times higher heat-resistance compared to ZrB_2-SiC. The low density and high heat-resistance properties make such ceramics promising for the manufacture of front edges of the nose fairing, the wing and stabilizer of hypersonic airplanes and reusable spacecrafts and hot components of jet engines of all types. It is possible to produce coatings of ultra-high-temperature ceramics for spacecraft parts made from carbon-carbon composites, in order to enhance their performance in oxidizing environments.

Ceramic-metal composite materials—cermets are materials that are heterogeneous compositions of one or several ceramic phases with metals or alloys, possessing a set of properties that combine the properties of the original components. Composites in which the properties of metals improve the ceramic phase belong to infra cermets, and compositions in which the properties of ceramics are improved by metals belong to ultra cermets.

In the above definition, the content of metal and ceramic components is not limited, since for various combinations it varies significantly depending on mutual solubility, although it is sometimes said that cermets contain 15%—85% (volume) of ceramic phases. Any metals or alloys can be used as part of cermets. The ceramic component can be based on oxides or oxygen-free compounds.

Non-metals in cermets give them the required performance properties—heat resistance, hardness and wear resistance. The metal component combines solid particles in a cermet in a single composite material, providing products with the necessary strength and ductility. Therefore, the properties of cermets depend on the properties of the matrix and the filler, the volume ratio and the adhesion between them.

The components of cermets must meet special requirements with regard to chemical stability, thermal compatibility and the ability to form bonds at the phase boundary. The requirement of chemical stability determines such combination of metal and ceramic phases so that chemical interaction does not occur between them with the formation of compounds or that the metal does not dissolve completely in the ceramic phase, which would turn the composition into a mixture of ceramic phases or a single-phase material[111].

Classical cermets are obtained by the technology of liquid-phase sintering, but they can also be obtained by solid-phase sintering of mixture of powders, including mixtures of non-metallic and metal-like refractory compounds. With approximate equality of the volume contents of the refractory metal and non-metal (oxide) phases, sintering is sometimes possible at temperature above the melting point of the oxide.

The main property of cermets, which determines their use in aircraft with turbojet engines, is their strength at high temperatures. So, the long-term strength of carbide cermets at high temperatures, despite the rather high metal content in them, is much higher than the long-term strength of heat-resistant metal alloys. Cermets are used for the manufacture of turbine blades and nozzle apparatus, flame stabilizers and injectors for afterburner chambers in direct-flow jet engines.

The specific thrust of the engines is directly dependent on the gas temperature at the turbine inlet, with all other things being equal. There is not a problem to obtain a high gas temperature; on the contrary, in practice, to reduce the temperature, excess air is supplied to the engine in order to ensure the performance of metal superalloys. The main structural components of the engine, limiting the temperature of the gas, are turbine blades. In the range of tensions, usually occurring in the blades, the use of cermets with the highest long-term durability allows to increase the gas temperature by 120 ℃, if we take into account their low density and introduce an appropriate correction (the main stresses arising in the parts influenced by centrifugal forces material). Thus, a simple replacement of high-temperature alloys with cermets can achieve a 40% increase in traction power.

In addition to high strength at high temperatures, aviation materials must have resistance to thermal shock, high impact strength and reliability in operation. Very often, when flying at high altitudes or switching from one fuel tank to another, combustion in the combustion chamber is interrupted. At this moment, hot blades of the turbine begin to be blown with high intensity by the outside air, whose temperature is at a high altitude of ~-55 ℃. This is the most dangerous moment in the operation of the blades, because with a sharp cooling in the surface layers there are significant tensile stresses, leading to the occurrence of cracks. With the usual engine start, the blades heat up to 700 ℃ in 6 s. This means that all parts exposed to hot gases must be heat resistant. A number of cermets meets this requirement. It should be noted that at room temperature, not all superalloys have the necessary resistance to thermal shock.

An important property of materials operating in a hot gas stream is impact strength. During operation, particles from outside often get into the turbine. Such particles are sometimes captured by the engine, and in some cases are formed in the engine itself, due to the destruction of some parts, such as the flame tube of the combustion chamber. At the same time, the most serious problem is caused by the formation of solid carbon particles in combustion chamber of certain types of engines. Since the edges of the blades move at a speed of several hundred meters per second, the impact of even small particles on the blades will cause damage to the engine, which can lead to a catastrophe.

The main method for producing Al_2O_3—metal cermets is slip casting, which provides the maximum variety of shapes (bevels, variable cross-sectional dimensions, V-shaped notches, etc.). The overall dimensions of the product are limited only by the strength of the briquette and the size of the thermal sintering equipment.

The idea of producing the Cr-Al_2O_3 cermets was to use a combination of Cr and Al_2O_3 to obtain a material that preserves the favorable properties of both components. As a result, a cermet with the desired high temperature properties, though insufficient ductility, was obtained. It should be noted that the lack of the ability to react elastically to shocks (insufficient impact strength) is a common shortcoming of all cermets. The chemical composition and physical properties of the cermet of Cr-Al_2O_3 type are presented in Tables 6.2 and 6.3.

TABLE 6.2 Chemical composition of Cr-Al$_2$O$_3$-based cermets

Cermet	Content of chemical elements/% (metal/ceramic=65%/35% of volume)				
	Cr	Al$_2$O$_3$	Mo	TiO$_2$	W
Cr-Al$_2$O$_3$	77	23	—	—	—
Cr-Mo-Al$_2$O$_3$-TiO$_2$	59	19	20	2	—
W-Cr-Al$_2$O$_3$	25	15	—	—	60

TABLE 6.3 Physical properties of Cr-Al$_2$O$_3$-based cermets

Characteristic		Cermet		
		Cr-Al$_2$O$_3$	Cr-Mo-Al$_2$O$_3$-TiO$_2$	W-Cr-Al$_2$O$_3$
ρ (at T=22 ℃)/(kg·m^{-3})		5 900	6 000	8 800
Oxidation (at T=1 204 ℃ per 100 h)/(mm·year^{-1})	continuous work	2.67	2.41	952
	periodic work	3.33	4.05	5 867
$\alpha \times 10^6$ (in the range of 25—1 000 ℃)/℃		8.93	8.48	8.36
λ (at T=260 ℃)/[W·(m·℃)$^{-1}$]		50	—	—
c (calculated)/[kJ·(kg·℃)$^{-1}$]		0.672	0.588	—

Cermets of Cr-Al$_2$O$_3$ type are resistant against oxidation to temperature of 1 200 ℃. This turned out to be possible due to the close-fitting of Cr$_2$O$_3$ layer formed on the chrome phase of the cermet when heated and the inert nature of Al$_2$O$_3$. Such cermets have some strength ($R_{m,\text{ben}}$ was measured by the method for non-plastic materials) at temperatures when many metals are already melting (see Table 6.4).

TABLE 6.4 Mechanical properties of Cr-Al$_2$O$_3$-based cermets[127]

Characteristic	Cermet			T/℃
	Cr-Al$_2$O$_3$	Cr-Mo-Al$_2$O$_3$-TiO$_2$	W-Cr-Al$_2$O$_3$	
HRC/MPa	270	500	520	—
$R_{m,\text{ben}}$/MPa	316	393	513	20
	190	380	471	982
	127	204	316	1 149
	32.4	83	176	1 316
E/GPa	264	267	267	20
	229	221	229	1 000

(continued)

Characteristic	Cermet			T/°C
	Cr-Al$_2$O$_3$	Cr-Mo-Al$_2$O$_3$-TiO$_2$	W-Cr-Al$_2$O$_3$	
R_m/MPa	148			20
	144			427
	141			649
	138	—	—	816
	118			982
	82			1 093
G/GPa	119	—	—	20
τ_{sh}/MPa	280	—	—	20
μ	0.20—0.22	0.25—0.27	—	—

Flame stabilizers are made from cermet Cr-Al$_2$O$_3$, designed to regulate the size and shape of the flame in the combustion chamber of a jet engine. Due to slip casting technology, such stabilizers are molded with small tolerance, which eliminate finish processing. If necessary, the exact dimensions of cermet Cr-Al$_2$O$_3$ can be processed by hard alloy technologies.

In cermets of Cr-Mo-Al$_2$O$_3$-TiO$_2$ type, by the addition of Mo and TiO$_2$ the hardness and strength of the material increase. There is a significant increase in flexural strength at all operating temperatures. Compared with Cr-Al$_2$O$_3$ cermet at temperature of 1 310 °C, this property is 2.6 times higher. Cermet of Cr-Mo-Al$_2$O$_3$-TiO$_2$ has higher resistance to impact, abrasive wear and sliding friction. The remaining properties are similar to Cr-Al$_2$O$_3$ cermet, however, the resistance to oxidation is slightly lower at temperatures above 1 100 °C. Cr-Mo-Al$_2$O$_3$-TiO$_2$ cermets are intended for manufacturing high-temperature seals and bearings in gas turbines[127].

In cermets of W-Cr-Al$_2$O$_3$ type, tungsten is the main component, which contributes to increase in hardness and a significant increase in strength. The flexural strength in bending $R_{m,ben}$ of such a cermet is higher than that of most high-temperature non-metallic materials. Under conditions of severe erosion and abrasive wear (for example, when testing a rocket nozzle), W-Cr-Al$_2$O$_3$ cermet keeps its original shape longer than most non-metallic materials. However, the low content of chromium compared with cermet Cr-Al$_2$O$_3$ and Cr-Mo-Al$_2$O$_3$-TiO$_2$ reduces its resistance to oxidation[127].

Inserting of a rocket nozzle from W-Cr-Al$_2$O$_3$ cermet was tested for resistance to erosion and wear at gases emerging from the rocket nozzle, reaching several thousand degrees. When the rocket engine is operating on solid fuel, regenerative cooling cannot be carried out. Therefore, the material of the inserts must have high thermal conductivity, heat capacity, strength at operating temperatures, resistance to oxidation and erosion in gas flows moving at supersonic speed. Tests have shown that the effect of erosion on W-Cr-Al_2O_3 cermet is significantly less than on graphite and materials based on SiB_4 and SiC[127].

The LT-2 cermet based on Al_2O_3, having $R_{m,ben}=176$ MPa at temperature of 1 320 ℃, was developed for structural parts of aircraft engines. *TiC-metal* cermets are known under the name "Kentiumiums". The material K163B1 (TiC-33Ni-7Mo) and WZ12c (TiC-33Ni-10Co-10Cr) positively withstand harsh conditions during operation of aircraft engines. The thermal conductivity of such materials is 2—3 times higher than that of austenitic steels and high-temperature alloys. Cermet K183A (TiC-32Ni-2.5Cr-3Mo-2.5Al) with $R_{100}^{980}=197$ MPa is designed to work in gas turbines[127].

Boride-metal cermets (ZrB_2-Mo, TiB_2-Mo, CrB_2-Mo, TiB_2-Co) are promising for manufacturing gas turbine blades, and have refractory eutectics with a melting point of 1 500—2 200 ℃[59].

Chapter 7 Materials for Friction Joints

7.1 Introduction

  Friction is an amazing phenomenon of nature. It gave humanity warmth and fire, the ability to stop moving in a short period time, speed up chemical reactions in 100 thousand times, hear the sound of a violin, create a car, a plane and much more.

  Till today, friction remains a mystery in many ways. Only during friction do the mechanical, electrical, thermal, vibration and chemical processes take place simultaneously. Friction can strengthen or weaken a metal, increase or decrease its carbon and hydrogen content, turn gold and platinum into oxides, polish metals and weld them. Friction is a self-organizing process, during which phenomena aiming at destroying the surface or, conversely, producing a whole series of systems that reduce wear from friction (in particular, the effect of wear-free friction) occur with a certain sequence and very "reasonably".

  Today, one of the most serious problems related to friction is the wear of machines and mechanics. The annual loss because of friction and wear in the United States is approximately 100 billion USD. A significant part (about 70%) of the world energy resources is spent on friction, and 70%—90% of mobile interfaces in machines fail due to wear. At the same time, efficiency, accuracy, reliability and durability are sharply reduced, and dynamic and acoustic characteristics are significantly deteriorated. In developed countries, only the loss of abrasive wear ranges from 1% to 4% of the national product.

  The development of engineering, in particular, mechanic engineering, has given great importance to the durability of machinery in terms of saving material resources and labor and has brought a wide range of designers, technologists, operators, and scientists of various specialties closer to those problems. This made it possible, based on the achievements in physics, chemistry, mechanics and materials science, to lay the foundations of the theory of friction, wear and lubrication of machines. Nevertheless, the state of the science of friction today does not allow us to calculate the wear parts for durability and predict their service life. Predicting the service life of friction pairs is based on qualitative dependencies on various friction machines or on bench devices simulating the operation of a particular joint obtained from the operation or experimental studies, which is very expensive and time-consuming. The science of tribology deals with problems of friction, and the tribotechnics is the applied part thereof.

Tribology is the science of contact interaction of solids with their mutual movement, which explains the problems of friction, wear and lubrication, and studies the processes of interaction of surfaces in their relative motion. The term was introduced in 1966 by the Englishman P. Dass. In recent years, the concept of tribology includes tribo-physics, tribo-chemistry and tribotechnics.

Tribotechnics includes the applied part of tribology, methods and means of ensuring the operability of friction joints. This is the science on the contact interaction of solids in their relative motion, covering the whole range of friction issues, wear and lubrication of machines.

7.1.1 Basic Terms of Tribology

Abrasive ability is the property of abrasives to destroy the surface of materials during mechanical interaction with them (abrasive machining, external friction, etc.).

Abrasive wear is the mechanical wear of materials as a result of cutting or scratching the solids or loose particles.

Adhesive interaction is the occurrence of a friction bond between the films on the surfaces of contact interaction with a clearly visible interface between the contacting bodies.

Antifriction is a comprehensive performance characteristic of friction material, which ensures normal operation under friction conditions.

Antifriction materials are friction materials used in bearing or guide assemblies (sliding bearings), whose friction coefficient is 0.01—0.1 with lubricant and 0.05—0.5 without lubricant.

Boundary friction is the friction of two solids, on the surface of which there is friction of the liquid layer, whose properties differ from the properties in the bulk.

Boundary lubricates (*lubricating film*) are layers formed as a result of the adsorption of polar hydrocarbon molecules on the surface of solids (metals) under the influence of a solid phase field.

Break-in is the process of changing the roughness of friction surfaces and the physical and mechanical characteristics of the surface layers of the material in the initial period of friction, which plays a role in the reduction of friction work, wear rate and stabilization of friction coefficient.

Caricature refers to penetration of abrasive particles into the friction surface.

Cavitation wear is mechanical wear during the movement of a solid relative to a liquid, in which gas bubbles collapse near the surface, resulting in local high pressure or high temperature.

Cavitation-erosion wear is a complex corrosion-mechanical process, due to the micro-impact of the liquid on the surface of the part.

Coefficient of friction is the ratio of friction forces to the normal force of the pressing body to each other. The coefficient of adhesion is the ratio of the incomplete friction force to the normal force (the highest static friction force).

Compatibility is the ability of two or more materials to jointly perform designed functions.

Counterbody is a component of a friction pair that works in conjunction with friction material.

Depletion is the result of wear, defined in arbitrary units. Depletion can be expressed in units of length, volume, mass, etc. The most widely used concepts are linear or weight (mg/km) wear.

Erosion wear is the mechanical wear as a result of the influence of the flow of liquid and/or gas.

External friction is a phenomenon of resistance to relative displacement, arising between two bodies, in the areas of contact of surfaces tangentially to them, accompanied by energy dissipation. There are frictions of static state, movement, sliding, rolling and rolling with sliding.

Fracture (friction surfaces) is the process of wear and deformation of the surfaces of contact interaction.

Fretting is a phenomenon that occurs in the contact of two bodies with their slight mutual displacement. There are:
- fretting corrosion;
- fretting fatigue;
- fretting wear.

Friction materials are designed to work in joints that transmit or dissipate kinetic energy (brakes, clutches, dampers, etc.)

Friction of movement is the friction of two bodies in relative motion.

Static Friction is the friction of two bodies during micromovements before the transition to relative displacement.

Friction operation is the energy transferred by the thermodynamic system (tribopair) to the external environment and to the materials of the friction pair components in the event of a change in the external parameters of the system.

Friction with lubricant is the friction of two bodies in the presence of any type of lubricant introduced on the friction surface.

Friction without lubricant is the friction of two bodies in the absence of any type of lubricant introduced on the friction surface.

Gradient of mechanical properties is a vector characterizing the change in the mechanical properties of a component of a friction pair by tuning the friction surface.

Maximum friction force is the static friction force, whose any excess leads to the appearance of movement.

Hydroabrasive (gas abrasive) wear is the wear due to the abrasive solid particles entrained in a gas or liquid flow.

Hydrogen wear is the process of destruction of a metal components of a friction pair due to the absorption of hydrogen.

Internal friction is a phenomenon of resistance to the relative movement of parts of the same body due to two different groups of phenomena—inelasticity and plastic deformation.

Liquid friction is the friction of two solids, separated by a layer of liquid, in which its bulk properties appear.

Lubricant is the material introduced onto the friction surface to reduce friction force and wear rate.

Lubricating refers to leading the lubricant to the friction surface.

Lubrication is the effect of a lubricant, as a result of which a friction force and/or wear rate decreases between two surfaces.

Performance capability refers to the state of the object in which it can perform the specified functions, while keeping the specified parameters within the limits established by the regulatory and technical documentation.

Preliminary shift is the relative micromovement of two solids in the process of friction at the transition boundary from the static state to relative motion.

PV criterion is a qualitative criterion for evaluating the operation of a friction pair: P is the pressure on the support, V is the sliding speed. Based on this criterion, friction joints are conventionally divided into light-loaded, medium-loaded, and high-loaded.

Relative wear resistance is the ratio of wear resistance of the test material to the wear resistance of the reference material under the same conditions.

Rolling friction is the friction of motion of two solid bodies, at which the speeds of bodies at the points of contact are same in magnitude and direction.

Scuffing is the process of appearance and development of friction surface damage due to the setting and transfer of materials.

Secondary structures are new phases (a thin-film structure) that spontaneously form during friction due to the interaction of the surface layers of solids, lubricants and the gaseous medium.

Seizure is a local connection of two solids due to adhesion occurring during friction.

Sliding friction is the friction of the motion of two bodies, at which the speeds of bodies at the points of contact are different in magnitude and direction.

Sliding speed is the difference between the relative velocities of bodies at the points of contact for friction.

Surface of friction is the surface of bodies involved in friction. It is characterized by macrogeometric, microgeometric, physical, mechanical, chemical and other properties.

Tearing is friction surface damage in the form of wide and deep furrows in the direction of sliding

Theory of friction (wear) is a system about the basic laws of the friction (wear) process, which is based on interaction mechanism of friction bodies.

Tribosystem is a complex tribodynamic system formed by the interaction of friction bodies, intermediate medium and parts of the surrounding space.

Wear products are particles of materials that are separated from the material during wear.

Wear rate is the ratio of wear to the time interval during which it occurred.

Wear speed is the ratio of wear to the friction path or the friction force.

Wear resistance is the property of a material to resist wear in certain friction conditions, as measured by the reciprocal of the wear rate or wear spead.

Wear-out is the process of destruction and separation of material from the surface of a solid and the accumulation of its residual deformation during friction, with the gradual change in the size and shape of the body.

Weight wear is the mass of a worn substance Δm removed from a unit of the nominal contact area A per unit friction path L. By the type of interaction of the surface and the state of the lubricating layer, friction is classified as the following cases:

① friction of physically clean surfaces;
② friction with boundary lubrication;
③ friction with hydrodynamic (gas-dynamic) lubrication;
④ friction with hydrostatic (gas-static) lubrication;
⑤ friction at contact which is hydrodynamic lubrication;
⑥ friction in the conditions of electromagnetic interactions.

7.1.2 Some Examples of the Influence of Friction on the Operation of Mechanics

1. Example 1

The severe wear of bronze with composition Cu-10Al-3Fe-1.5Mn results in premature failure of the upper axle box of the shock absorber of the chassis after 400—500 landings. In order to reduce the effective loads, its height was increased to 1.5 times, but it did not eliminate the increased wear. Research has shown that the wear resistance of bronze is very low when lubricated with a nitroglycerin mixture (the shock absorber is filled so that it does not freeze in the air). Replacing this mixture with the hydraulic fluid AMG-10 (based on kerosene) eliminated the high wear of axle boxes.

2. Example 2

The axle of the wheels of the main landing gear broke on the plane "Ruslan" during the time of 2 000 landings at the airport of Amsterdam. The axle was made of high-strength structural steel Fe-0.3C-1Cr-1Mn-1Si-2Ni-1Mo. The fracture analysis showed that the damage is intergranular and did not occur in the zone of maximum stresses. The location of the destruction center coincided with the contact region of the steel axle, the aluminum spacer and the titanium brake housing. This connection of metals, along with the emergence of an electrochemical pair under vibration conditions, led to the appearance of fretting between the aluminum sleeve and the steel axle. Since alumina with high hardness was formed during fretting (as well as under friction), wear during micro-displacements mainly occurred in the steel part, which led to the formation of pitting. The design flaw—the shape of the spacer sleeve resulted in the accumulation of moisture in the place of pitting formation, leading to a further crack growth mechanism—stress corrosion. The changes to the design were as follows: adding a steel insert to the aluminum expansion sleeve, which excluded fretting—aluminum-steel contact, and removing the moisture at stagnant zones. The above mentioned design change made it possible to completely eliminate axle failures at the given location.

3. Example 3

On the Yak-42 aircraft, the screw nut was made of aluminum bronze (Cu-10Al-4Fe-4Ni) in the rudder control system. Despite the fact that the screw-nut design is generally standard, and bronze is a classic antifriction material, nevertheless, due to high specific loads and a rather "large" coefficient of friction 0.15—0.2 per pair of nitrated steel (screw), large stresses occurred for bronze with the absence of sufficient lubrication, which led to the failure of the thrust control and, as a result, the fall of one of the first Yak-42 aircraft. Replacing the bronze nut with BFG-50M material ($\mu=0.08$) made it possible to exclude such cases in a further operation. This is an interesting example showing that disregard for friction joints can lead to unexpected consequences even in non-friction joints.

These examples show that the increased wear of parts in the friction joints in some cases destroys the tightness of the working space. In others cases, the normal lubrication mode leads to loss of kinematic accuracy of mechanics, which adversely affects the controllability of the aircraft and its reliability. Wear and damage of the surface dramatically reduce the fatigue properties of metals, and may result in the destruction of parts, even under loads much lower than the calculated ones. Increased wear disrupts the normal kinematic interaction of parts in the joints, resulting in significant additional loads, vibrations and shocks in the joints, which can lead to emergency situations. In this regard, materials used in friction joints are among the most important materials, not only in aircraft manufacturing, but also in engineering in general.

Friction plays an important role in modern technology. The following attentions should be paid to the basic methods of controlling the performance of tribopair in the responsible aircraft joints:

① right choice, taking into account the actual operating conditions, the existing tribomaterials for friction and wear resistance, and the creation of new, more advanced materials;

② increase of wear resistance of friction surfaces by various methods;

③ rational design of mobile interfaces, improvement of their structural appearance;

④ improvement of lubricants and machine lubrication systems.

The task of sliding materials is to absorb the load for some time during a relative movement with minimal losses due to friction.

If two real bodies are brought into contact, then at least three contact points are initially formed. When applying force or due to their own body weight, the points of contact will begin to deform (the deformation is elastic or plastic). The subsequent contact points will appear, which will develop into contact pads. If we imagine that two solid bodies begin to move relative to each other, the following phenomena are possible:

① The available asperities on the surface are removed or crushed—mechanical theory;

② At close contact of the asperities and accompanied heating, welding contacts are formed with their subsequent destruction—the mechanical and physical theory;

③ Chemical reaction of surface oxidation occurs due to the heat generated. Cracking and exfoliation of oxide films is a mechanical-chemical theory.

Based on the above concepts, wear can be:

① oxidative—normal;

② adhesive wear—the formation and destruction of welding bridges;

③ abrasive;

④ erosion.

The working conditions of the friction joints are characterized by a large variety of sliding speeds, loads, lubricant compositions and the environment. There are also various requirements for the properties of materials from which tribological engineering parts are made of—low or high values of the friction coefficient, acceptable values for the wear rate, corrosion resistance, heat resistance, etc. Moreover, there is a complex of properties required that is not inherent to any metal or non-metal individual materials. This problem is solved by producing composite materials in which the individual components perform specific functions[72].

Very often in industrial technology, and in aviation particularly, there is a need to fabricate tribopairs that operate without lubrication, and those are cases when it is not possible to supply lubricants to the assembled joint ("without maintenance"). Such joints are produced by powder metallurgy methods from composite materials containing lubricants inside themselves, as it will be discussed below.

Materials for tribomechanical purpose are divided into bearing materials and frictional materials. The task of the materials is to convert mechanical energy into thermal energy, but in the first case this transformation should be minimal and in the second one, it should be maximum[5].

7.2 Bearing Powder Materials

Bearing materials are friction materials used for working in bearing or guide assemblies, whose friction coefficient is 0.05—0.1 with lubricant and 0.05—0.5 without it[119]. Bearing is the ability of a material to provide low coefficient of friction, low friction cost and low wear rate of the parts in contact. In addition, materials in specific operating conditions should be, for example, resistant to oxidation at elevated temperatures, have high resistance to corrosion when working in corrosive environments, etc. These materials should work in a wide range of sliding speeds (from 0.001 m/s to 100 m/s) and loads (from 1 MPa to 25 MPa), while temperatures are up to 100 ℃ and higher.

The concept of *"bearing"* includes the following properties:

① a good initial running-in, that is the time required to reduce and stabilize the coefficient of friction to a given value, should be minimal;

② high tribological properties—low coefficient of friction, low wear in a wide range of operating loads and speeds, etc. ;

③ the ability to withstand the load, speed and temperature without breaking down or changing shape and quality;

④ the ability to form self-lubricating or easily rubbed products of abrasion of colloids (film);

⑤ lower hardness than that of the shaft neck, and the hardness should not decrease in a certain temperature range;

⑥ high thermal conductivity and low coefficient of thermal expansion;

⑦ sufficient fatigue strength under dynamic loads and sufficient ductility to avoid brittle fracture;

⑧ good technological properties;

⑨ microporosity or microcapillary, contributing to sustaining lubricant on the rubbing surface;

⑩ good anti-corrosion properties;

⑪ sufficient static and dynamic strength at elevated temperatures;

⑫ processability and efficiency.

Bearing materials that satisfy all this complex of properties does not exist. Each bearing material shows its optimal performance in a certain range of operating parameters—the actual load, moving speed, lubrication conditions and chemical composition of the lubricant, environmental aggressiveness, including the presence of dust, vacuum, etc. The bearing of a particular material is judged by its coefficient of friction with a given counter body and lubrication conditions, and by the presence and extent of damage to friction surfaces, etc.

The use of bearing materials as intermediate inserts (bearings) is the most optimal design solution to ensure the friction unit performance—when worn, the bearing is removed and replaced with a new one.

For such purposes, the most effective materials are those produced by powder metallurgy methods that contain lubricant within themselves in their structure. The porosity of such materials is 15%—27%, and 98%—99% of the pores form an interconnected porous structure. Depending on the powders selected for the manufacture of the bearing, the pore diameter varies from 1—20 μm. 500 to 2 000 pores are provided every square millimeter surface. Interconnected pores can be built in a length of 1—2 km in 1 cm³ of bearing material. At least 90% of open pores are saturated with lubricant[128].

In porous oil-impregnated bearings, a lubricating film is formed, which is different from in solid bearings, where the lubricant is added externally. In solid (non-porous) bearings, a lubricating layer is formed at hydrodynamic pressure. This pressure results from the relative movement in the wedge-shaped gap between the shaft and the bearing. It reaches the required value for squeezing a pair of shaft sleeves apart, only at a certain speed of rotation. In contrast, in porous oil-impregnated bearings, even at start-up, there is an equilibrium between capillary forces in the bearing gap and in the pores. Thus, between the shaft and the bearing, there is a thin layer of lubricant, which is effective even at the beginning of the movement.

Thus, porous oil-impregnated bearings have good starting characteristics (see Fig. 7.1(a)). During rotation, the lubricating wedge, which is under pressure, begins to form as a result of the shaft movement. The pressure distribution of this wedge is shown in Fig. 7.1(b). The pressurized part of the bearing is completely filled with lubricant. In this case, a continuous and stable lubricating film is formed. The decrease in pressure in the pores of the bearing materials leads to the circulation of lubricant from the hermetic lubricating wedge in the low-pressure area. When stopped, the lubricant is absorbed into the pores due to the capillary effect, leaving only a thin residual film (see Fig. 7.1(c)).

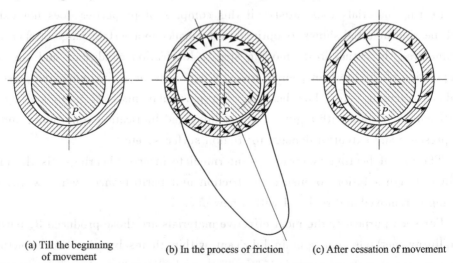

(a) Till the beginning of movement　　(b) In the process of friction　　(c) After cessation of movement

FIGURE 7.1　The work of porous bearings made by powder metallurgy. The distribution of forces acting on the structure and Lubrication[128]

7.2.1 Iron-based Bearing Materials

1. Iron graphite materials

Iron graphite materials are sintered iron-based bearing materials with a content of 1.5% to 10% graphite. Small additives of graphite (up to 1.5%) allow using iron graphite in unloaded units operating in the presence of a lubricant or in self-lubrication mode. In the latter case, their service life and allowable loads are much lower than when applying additional lubrication. When the content of graphite is more than 1.5%, with the technological support of the combination of fixed carbon and structurally free graphite, it is possible to improve the bearing properties of the material, especially in the presence of lubricant[129-130].

Iron graphite materials have a pearlite-ferritic structure. The amount of pearlite component depends on the initial content of graphite and sintering conditions. Pearlite structure has the greatest wear resistance. The content of the ferritic component is allowed up to 50% and depends on the operating conditions of the bearing joint[130].

Bearing properties are largely determined by the working conditions—load, slip speed and the presence of a lubricant. Under fluid friction conditions, the friction coefficient is in the range of 0.005—0.09, with boundary lubrication—0.02—0.125. The quality criterion is $PV \leqslant 1.6$ MPa · m/s without lubrication; $PV \leqslant 2.5$ MPa · m/s with boundary lubrication; $PV \leqslant 10$ MPa · m/s with fluid friction and $PV \leqslant 40$ MPa · m/s with pressure lubrication[129]. The maximum sliding speed for iron-graphite products is $v = 2$—3 m/s. At higher speeds, the temperature in the friction unit exceeds the allowable (100—130 ℃) range. The service life of iron-bearing bearings is 3 000—5 000 h[130].

Sintered iron graphite bearings come close to gray cast irons in their tribomechanical properties, but have better running power, smooth sliding and noiselessness.

Iron graphite materials are obtained by cold pressing at a pressure of 600—800 MPa to obtain green compacts with a given porosity of $p = 15\%$—20% and subsequent sintering in a protective (better reducing) medium or vacuum at a temperature of $T_{\sin} = 1\,000$—$1\,150$ ℃[59]. The properties of iron graphite products for aviation equipments are presented in Table 7.1.

2. Metal-glass materials

Metal-glass materials are composite materials consisting of one or several metals and glass phases. Products made from these materials are obtained by cold pressing of the prepared mixture with subsequent sintering of the blanks, hot static pressing of the blanks, and impregnation of porous blanks with glass melts.

TABLE 7.1 Properties of sintered bearing iron graphite

Graphite content/%	Porosity/%	HB/MPa	R_m/MPa	Maximum allowable	
				P/MPa	$v/(m \cdot s^{-1})$
2.5—3.0	17—35	300—1 450	70—280	3—10	4
7.0	18—25	250—750	150—190	7	4

During sintering of the blanks, free silica and metal oxides interact at the interface of the solid metal and liquid glass phases. Molten glass well wets all metals in the presence of an oxide film on their surface when they reach the temperature sufficient for the interaction of oxides with silica melt. This temperature is usually below the melting point of the metal matrix but must significantly exceed the temperature of the melt. When the softened glass connects the surface of the oxidized metal, the oxides dissolve in the glass, forming complex compounds (silicates).

The presence of the glass phase in the material activates sintering processes and contributes to increasing the density of the material, increasing the number of "closed" pores and reducing the stress concentration.

Metal-glass materials are characterized by high wear resistance, corrosion resistance and durability, which are determined by a combination of a soft metal base and solid glass inclusions (Charpy's rule). Metal-glass bearing materials based on carbon steel of eutectoid composition with 15% of glass powders have proven to work in harsh conditions of space, where friction joints work in vacuum, the inability to use liquid lubricants, large temperature differences, solar radiation and high contact loads on tribojoints. The production technology of such materials consists of pressing a mixture of metal and glass powders, sintering, hot static pressing and heat treatment (quenching with tempering). At the same time, the following materials properties are obtained[131-132]. R_m = 580—620 MPa; $R_{m,ben}$ = 1 000—1 200 MPa, A=2%—3%, HRC=52—58.

7.2.2 Copper-based Bearing Materials

Bronze-graphite composite is sintered bearing materials based on copper, containing 6%—12% of tin and 1%—25% of graphite. The graphite present in bronze is always in a free form, which gradually forms a graphite film on the surface of the counter body in the process of friction, which is constantly restored after mechanical damage in certain areas of the friction surface (under optimal friction conditions). Bronze-graphite materials work more reliably than materials without graphite and are increasingly replacing bronze and brass in the friction joints of various mechanics especially in aircraft construction.

The magnitude of the permissible pressures for bronze-graphite composite largely depends on the porosity of the material and the sliding speed. With an increase in porosity and sliding speed, the load-carrying capacity of the bronze-graphite composite decreases.

Increasing the content of graphite above 6% dramatically reduces the strength of the material. The effect of graphite on the bearing properties of bronzes is very specific—the lower the graphite content is, the more suitable it is for working under high loads, and at the same time, a material with a high content of graphite is much more suitable for working at high slip speeds with low loads.

The simplest version of such materials is porous bronze (Cu-10Sn-4C) containing graphite impregnated with oil as a solid lubricant. This is one of the first powder bearing materials which have been applied to lightly loaded friction joints in aircraft construction for many decades. The advantages of powder oil-impregnated bearings with compared to cast are:

① high running-in smoothness and noiseless operation;

② possibility to be used in the pendulum motion and rotation with velocities lower than those necessary for the formation of a hydrodynamic wedge;

③ possibility to be installed in a vertical or inclined position;

④ reduction of oil consumption;

⑤ reduction of wear of a friction pair, including the reduction of wear due to a decrease in starting loads.

The disadvantages of powder oil-impregnated bearings include:

① limited load and sliding speed due to low strength compared to cast materials;

② high sensitivity to shocks and pressure on the edge due to the presence of porosity[35].

In aircraft manufacturing, porous bronze is used for producing slide bearings, which operate in self-lubrication mode with operating parameters—$P=4-5$ MPa, $v \leqslant 1$ m/s.

Fig. 7.2 shows a typical technological scheme for manufacturing the bronze-graphite bearings. The microstructure of such bearings consists of Cu crystals with solute Sn together with graphite and pores which evenly distribute over the thin section. The effect of self-lubrication is ensured by both graphite and lubricant absorbed in pores in an amount of at least 1%[5]. Bearings made from bronze-graphite composite are shown in Fig. 7.3.

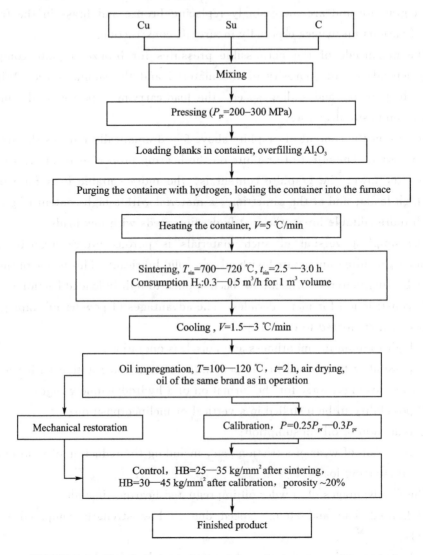

FIGURE 7.2 Manufacturing technology of powder bronze-graphite products

For parts operating at elevated loads and temperatures, the use of liquid lubricants is impractical because of their blocking or squeezing from the friction surface due to a decrease in viscosity. For such purposes, complex-alloyed materials based on copper or nickel are used, which are alloyed with Ti, Pb, Co, Zn, Fe, Al to increase their strength properties and introduce substances that play the role of solid lubricants, for example, MoS_2.

Examples of such materials are complex-alloyed bronze (Cu-17Ni-6Sn-4Pb-5MoS_2-5C), from which bearings are used in airplanes for work in "without

(a) Porous oil-impregnated bronze-graphite bearings   (b) Complex alloyed bronze bearings

FIGURE 7.3 Powder bearing bearings

maintenance" conditions at temperatures of up to 350 ℃ (see Fig. 7.3). The material containing up to 10% of solid lubrication of a layered structure (C, MoS_2), which reduces the coefficient of friction, helps to reduce wear and the stuck likelihood with a mating part. Such materials are capable of operating in "dry" friction units in an aircraft with $P \leqslant 5$ MPa and $v = 0.5$ m/s without lubrication in combination with steels with a friction coefficient of 0.20—0.24[5].

Such components are made based on the technology similar to that shown in Fig. 7.2 with some changes: mixing of the initial powders, pressing at a pressure of $P = 350$—500 MPa, sintering in a medium of dry hydrogen at a temperature of $T = 820$—850 ℃, calibration at $P = 500$—650 MPa with subsequent annealing at $T = 810$—820 ℃. The mechanical properties of the bronze at room and elevated temperature are given in Table 7.2[5].

TABLE 7.2 Mechanical properties of sintered complex-alloyed bronze ($Cu-17Ni-6Sn-4Pb-5MoS_2-5C$)

Material condition	Test temperature/℃	$R_{m,ben}$/MPa	$R_{m,com}$/MPa	τ_{sh}/MPa	HB/MPa	KCU/ (kJ·m^{-2})
Sintered	20	130—140	250—280	120—140	550—750	19.6
	350	70—90	160—180	80—90	400—650	14.7

7.2.3 Metal-polymer Bearing Materials

For operation in heavily loaded friction joints, for example screw-nut tribopair in the control system of the aerodynamic surfaces of the wing, *metal-polymer materials* are used. The most common metal-polymer bearing material is a Metal-Fluoroplastic Tape (MFT), which has a three-layer composition (see Fig. 7.4):

1st layer—bearing basis of low carbon steel with 0.75—2.25 mm thickness, with a layer of copper deposited by electroplating;

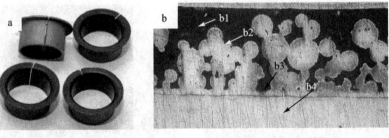

(a) Collapsed sleeves　　　　　　(b) Microstructure (×100)

b1—Teflon with MoS$_2$ filler; b2—Spherical sintered bronze granules;

b3—Copper layer; b4—Steel basis[133]

FIGURE 7.4　Metal-fluoroplastic bearings

2nd layer—porous (porosity ~30%), 0.35 ± 0.05 mm thick, consisting of spherical tin bronze particles with a dispersity of 80—120 μm (63—160 μm allowed) layer, baked on one side to the copper-plated steel basis;

3rd layer—a bearing layer consisting of a fluoroplastic filled with molybdenum disulfide, filling the pores between the bronze particles.

This design of composite bearing material is dictated by the following considerations:

① The steel basis acts as a supporting structure and removes heat from the friction zone;

② Copper plating is performed to improve the sintering conditions of the porous layer;

③ The porous layer of bronze performs two functions: participating in friction as the bearing material and acting as reservoir for the fluoroplastic;

④ The fluoroplastic layer is the main antifriction material, which has the lowest coefficient of friction among all solid materials (μ=0.05—0.08), but has a significant drawback — cold flow under load (cold flow); therefore, it is filled with disulfide or graphite.

When such material is in use, the temperature rises in the contact zone, which leads to the displacement of the fluoroplastic from the pores to the friction surface due to the difference in the thermal expansion coefficients of bronze and fluoroplastic. In this case, the fluoroplastic forms a film on metal surfaces, which reduces the friction coefficient and, as a result, the temperature in the contact zone.

Several thousand pieces of metal fluoroplastic bearings with various sizes are used in one plane. The materials are operable in the temperature range T = (−60)—(+250) ℃, depending on the counter body and it has μ = 0.06—

0.12 (without lubrication) at $P \leqslant 100$ MPa and $v \leqslant 0.1$ m/s. Often sleeves of metal-fluoroplastic tape (see Fig. 7.4(a)) work in aircraft structures during reciprocating motion under the action of a constant load in a cyclic mode. The work cycle of the sleeve in a specific design is the following: load application ($P = 100$ MPa at $T \leqslant 600$ ℃; $P = 50$ MPa at $T = 60—250$ ℃), making the shaft move at a given angle from one extreme position to another and back (deviation angle $\pm 30°$, $v \leqslant 0.04$ m/s—average per cycle). Sleeves must withstand 10^5 cycles in this mode.

A variation of the metal-fluoroplastic composite material is a material called BFG - 50M, which is a composition consisting of a basis (Cu-7Sn-0.2P tin bronze), a porous layer of the same bronze with thickness 0.5 mm and porosity 45%—50%, as well as an antifriction layer of fluoroplastic filled with graphite. The processes for manufacturing BFG - 50M are shown in Fig. 7.5 with details. The self-lubricating material BFG - 50M is designed to operate at $P \leqslant 100$ MPa and $v \leqslant 1$ m/s in a wide range of temperatures $T = (-60)—(+260)$ ℃. The friction coefficient of the material $\mu = 0.06—0.09$.

The peculiarity of this material is that bronze is not only the bearing basis of the bearing, but also a structural material. Initially, the material was developed with a specific design—a nut in the nut-screw mechanism of flap control (instead of ball-screw mechanism), which has the effect of self-locking in case of transmission failure. That is, the friction surface must be of a complex configuration—a trapezoidal thread with exact geometrical dimensions to ensure operability (see Fig. 7.6). The use of such nuts reduced the weight and dimensions of the sliding friction joints in the design of the aircraft screw mechanisms, as well as eliminated their lubrication and shortened the time of routine work and pre-flight preparation[5].

In addition to parts with a helical friction surface, the BFG - 50M material is also used in other friction joints of aviation technology—inserts for steering columns, running rollers of cargo doors, sliding bearings in the wing mechanization control system, etc.

Slip friction joints with parts from BFG - 50M provide an additional advantage when operating airplanes on unpaved airfields, where it is difficult to avoid the ingress of abrasive particles into the friction zone. Abrasive particles are embedded in the porous layer of material and are coated with fluoroplastic, which does not affect the operation of the friction joint as a whole[124].

The material BFG - 50M has a number of advantages as compared to the metal-fluoroplastic tape—the porosity of the non-spherical bronze layer increased up to

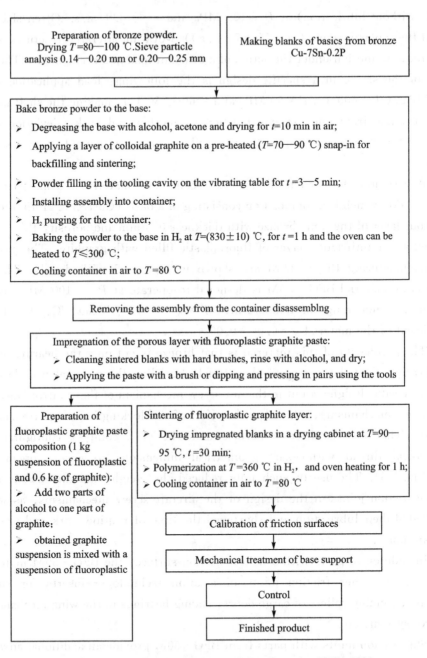

FIGURE 7.5　Technological scheme of manufacturing BFG－50M

45%, which allows increasing the amount of fluoroplastic as the main antifriction material and increasing the wear resistance; significantly higher corrosion resistance and bearing capacity of the basis; ability to manufacture parts of complex geometric shape with exact dimensions. Examples of friction bearings made from BFG－50M material are shown in Fig. 7.7. Using the manufacturing technology of the

 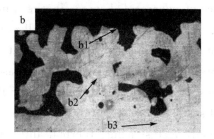

(a) The appearance of the nut   (b) The microstructure of the antifriction layer (×100)

b1—Fluoroplastic layer; b2—Non-spherical bronze granules; b3—Bronze base

FIGURE 7.6 Nuts from the material BFG-50M

material BFG-50M, it is possible to obtain parts with high dimensional accuracy without using mechanical processing at the final stages. Obtaining the parts of the tribopair with exact dimensions leads to a significant increase in their reliability and service life.

FIGURE 7.7 Friction bearings made from BFG-50M material

7.2.4 Nickel-based Bearing Materials

Sintered bearing materials based on nickel are used in aircraft manufacturing for producing the friction joints operating under severe conditions, in particular at elevated temperatures. In its pure form, nickel has good ductility, heat resistance, corrosion resistance, but poor antifriction properties when working without lubrication. To combine the positive properties of nickel with improved tribomechanical characteristics for the manufacture of high-temperature bearing products the powder alloys are used. Nickel, as a basis for providing temperature characteristics, adds hardening and bearing additives to obtain materials with a combination of high performance characteristics. To improve economic performance, nickel is partially replaced by iron and copper, whose content should not lower the corrosion characteristics.

When flying in clouds, an icing-up of an airplane often occurs due to the presence of supercooled water in the atmosphere at negative temperatures. The surface of the aircraft becomes crystallization centers of water and is covered with a layer of ice. At the same time, the flight characteristics of the aircraft significantly deteriorate (the vertical climb rate decreases, the maximum flight speed decreases, fuel consumption and power required for flying at a given speed increase), and its control systems may be jammed, which leads to disasters in practice. To protect the aircraft against icing, thermal anti-icing systems have been developed, whose performance is based on the heating of the surfaces to be protected with hot air. In order to compensate for the change in the length of the pipeline transporting the hot air with significant temperature drops, special compensators, which allow linear and angular movements, are in use. In order to minimize the stresses that occur when the componenting of the compensator move with the pipeline lengthening, these parts should be made from materials with minimum coefficient of friction over a wide temperature range (air from the engine is at a temperature of 500—600 ℃). In order to avoid depressurization of the compensator, the materials of the anti-friction insert must have low wear during operation. For this purpose, bearings made from nickel powder alloy Ni-21Cu-11Fe-1.5Sn-2.4Pb-4.5C-5MoS$_2$ are used. Such material (see Fig. 7.8) has working temperature of up to 600 ℃, friction coefficient $\mu=0.20$—0.24 when working in pairs with Fe-18Cr-10Ni steel without lubrication at temperature $T=550$ ℃ and sliding conditions $v=0.1$—0.2 m/s, $P=1.5$ MPa. The mechanical properties of the nickel-based bearing material at different temperatures are given in Table 7.3.

FIGURE 7.8 Blanks of bearing insert of compensators of anti-icing system of the aircraft from nickel alloy[124]

TABLE 7.3 Mechanical properties of Ni-21Cu-11Fe-1.5Sn-2.4Pb-4.5C-5MoS$_2$ sintered nickel alloy[5]

Material condition	Test temperature/℃	$R_{m,\text{ben}}$/MPa	$R_{m,\text{com}}$/MPa	HB/MPa
Sintered	20	180—230	300—340	600—680
Calibrated	450	120—170	200—270	—
Annealed	600	50—70	130—180	—

Parts made from Ni-21Cu-11Fe-1.5Sn-2.4Pb-4.5C-5MoS$_2$ alloy are produced by pressing a mixture of powders at a pressure of $P=500—550$ MPa and sintering in hydrogen medium at a temperature of $T=900—910$ ℃. Calibration is carried out at a pressure of $P=650—670$ MPa, followed by annealing in hydrogen at a temperature of $T=890—900$ ℃[124, 128].

7.3 Friction Powder Materials

Friction materials are designed to work in dynamic devices that transmit or dissipate kinetic energy (brakes, clutches, variators and others). These devices are designed to accelerate (clutch) or stop (brakes) the movement, as well as to transfer and change the direction of motion (regulators, variators, various friction clutches).

Friction materials operate under severe wear conditions at high specific loads and sliding speeds (up to 50 m/s), which lead to an instantaneous rise in temperature in joints. When an aircraft is landing, the brakes in the chassis heat up to more than 1 000 ℃.

The mechanical energy of moving elements turns into heat, and then dissipates when braking. Thermal exposure in parallel with reusable cyclic loading from both external exposure and abrupt thermal changes imposes the following special requirements on friction materials:

① high frictional heat resistance, that is, the ability of a friction pair to maintain a stable high coefficient of friction and to have low wear in a wide temperature range.

② high and stable friction coefficient. The coefficient of friction materials is in the range of $\mu=0.2—0.5$, and the process of friction should be stable without jerks.

③ fast running-in ability.

④ resistance to hardening. Friction pairs should not be stuck or even welded both in the process of braking and after it when the hot brake is in the static state.

⑤ thermal fatigue resistance.
⑥ high corrosion resistance.
⑦ fire safety. When braking, friction materials and wear products should neither ignite, nor to give off smoke and unpleasant odors.
⑧ sufficient mechanical strength.
⑨ wear resistance.
⑩ high thermal conductivity and heat capacity.
⑪ smoothness and noiseless engagement and slip.
⑫ manufacturability.
⑬ profitability.

High and versatile requirements for friction materials can be met only by creating complex composites. The use of sintered friction materials allows increasing the durability, reliability and efficiency of friction joints. Such materials not only improve the technical characteristics of aviation technology, but also provide high economic efficiency by increasing the durability of friction units and reducing operating costs. In addition, the safe operation of vehicles and the lives of people depend on the reliable operation of friction joints of aircraft.

Friction elements with sintered materials for aeronautical engineering are made in the form of discs, sector pads, and pads of various configurations. As a rule, they are a structure consisting of a steel support frame, lined on one or both sides with a layer of sintered friction material. The efficiency of the structure is determined by the properties of the friction layer. The strength properties, wear resistance and heat resistance of sintered friction materials largely depend on their structure, physical and chemical properties of the base. The metal base of friction materials is copper or iron alloys. To optimize the tribomechanical properties, the experts introduce various metallic and non-metallic additives, which, depending on the effect, can be divided into two groups:

① materials that reduce the tendency of the rubbing pair to jam;
② materials creating and stabilizing a certain friction mode.

To improve the anti-scuff properties and increase the wear resistance of friction materials, low-melting metals (Pb, Bi, Sb) and non-metallic substances (C, MoS_2, BN) are introduced into their composition. Fusible metals are introduced to "self-regulate" the friction coefficient. When the melting point of the low-melting component is reached, a liquid metal film is formed on the surface of the friction pair, which reduces the friction coefficient and, hence, heat generation. The formation of a liquid lubricating film contributes to a smooth and stable slip, which is especially important at elevated temperatures when the metal matrix has a greater tendency to setting and sticking.

The introduction of solid lubricants (metallic and non-metallic) reduces the wear of materials, and contributes to the stable operation of friction pairs, however, this reduces the coefficient of friction. To increase the friction coefficient to the required level and prevent the friction materials from setting, friction additives are introduced. The main task of friction additives is to ensure an optimal level of engagement with the working surface of the counter body without abrasive wear of the contact part.

For materials based on copper and iron, effective friction additives that increase the friction coefficient and resist to the setting, especially at high temperatures are non-metallic components—SiO_2, MoO_3, SiN, asbestos, and others.

Sintered materials of friction destination were first proposed in 1929, specifically for the needs of aviation technology.

In aviation technology, there are four main classes of friction composite materials-materials based on copper, based on iron, cermets and carbon-carbon friction materials.

7.3.1 Copper-based Friction Materials

Copper-based friction materials are used in the joints operating with the presence of lubricant and without it—electromagnetic and safety couplings of aircraft joints(see Fig. 7.9(a)). These materials have low wear and do not wear out the counter body. They have a stable friction coefficient ($\mu=0.28$—0.35), and the maximum operating temperature is 400 ℃. The most common materials of the compositions are Cu-8Sn-3SiO_2-6Pb-3MoO_3-3C and Cu-15Fe-11Sn-2Mn-3MoS_2-10C-7SiO_2-2SiC-2Al_2O_3. Such materials are made by pressing the mixture at a pressure of $P=250$—450 MPa and sintering under pressure in H_2 at a temperature of 750—800 ℃. The mechanical properties of the friction alloy Cu-8Sn-3SiO_2-6Pb-3MoO_3-3C at room temperature are given in Table 7.4.

TABLE 7.4 Mechanical properties of Ni-21Cu-11Fe-1.5Sn-2.4Pb-4.5C-5MoS_2 friction metal ceramics[5]

Material condition	R_m/MPa	$R_{m,ben}$/MPa	$R_{m,com}$/MPa	τ_{sh}/MPa	HB/MPa	KCU/(kJ·m^{-2})
Sintered	50—60	70—80	140—150	70—80	600	9.8—14.7

The tribological characteristics of such material are $\mu=0.28$—0.30 paired with steel at $P=30$ MPa and $v=0.1$—6.5 m/s, and the wear of such materials is 8 μm/h.

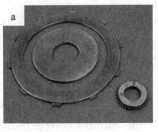
(a) On a copper basis

(b) On an iron basis

FIGURE 7.9 Products from friction powder alloys

7.3.2 Iron-based Friction Materials

Sintered friction materials based on iron have greater strength and can withstand heavy loads and friction temperature in comparison with copper, therefore they are used to work in more severe friction conditions—in wheel brakes, in aircraft wing mechanization systems, on-board loading cranes and other devices (see Fig. 7.9(b)). These materials, however, have more wear comparing with the counter body and have a less smooth working stroke than copper. They have a friction coefficient in the range of $\mu = 0.25$—0.4, and the maximum working temperature is 1 200 ℃. The chemical composition of the most common aviation friction materials is given in Table 7.5.

TABLE 7.5 Chemical composition of iron-based friction metal ceramics

wt. %

Grade	Fe	Cu	C	S	Si	Ni	Cr	W	B	Asbestos
FMK8	basis	5—7	5.5—6.6	1.3—2.3	—	23—27	8.5—10.5	5—7	—	—
FMK11	basis	14—16	6.5—8	0.5—1	1—2	—	—	—	—	2—3.5
FMK79	basis	9—11	6—8	0.2—1	2—3	—	—	—	1.5—3	—
MVK50A	basis	9—11	8.5—11	0.6—1.2	3—4.3	—	—	—	3—4.3	2.5—4

The physical and mechanical properties of friction materials based on iron are given in Table 7.6.

TABLE 7.6 Physical and mechanical properties of sintered friction materials based on iron

Indicator	Material grade			
	FMK8	FMK11	FMK79	MVK50A
$\rho/(kg \cdot m^{-3})$	7 000	6 000	5 500—5 600	5 000
$p/\%$	—	5—10	5—10	5—10

(continued)

Indicator	Material grade			
	FMK8	FMK11	FMK79	MVK50A
R_m/MPa	90—100	55—56	40—50	30—40
$R_{m,com}$/MPa	450—500	300—350	250—260	155—210
$R_{m,ben}$/MPa	—	160—200	110—120	100—140
τ_{sh}/MPa	80	80—100	90—110	67—85
KCU/(kJ·m^{-2})	—	—	9.8—10.8	7.85—11.77
HRF	630—830	700—950	800—1 050	600—1 000
$\alpha \times 10^6$ (in the range of $T=20$—600 ℃)/℃	—	13.3—14.8	14.3	10.9—11.3
c (at $T=600$ ℃)/[kJ·(kg·℃)$^{-1}$]	—	0.503—0.711	0.798	0.503—0.837

The friction properties of metal ceramics on an iron basis are given in Table 7.7.

TABLE 7.7 Tribomechanical characteristics of iron-based friction materials paired with cast iron and steel

Friction pair	μ	Stability (Δf)	Wear during 1 braking/μm	
			Metal ceramics	Counter body
FMK8+cast iron	0.16—0.2	0.65	16	4
FMK11+cast iron	0.25—0.29	0.77	22	4
FMK11+steel	0.18—0.22	0.80	24	6
FMK79+cast iron	0.3—0.4	0.70	12	4
MVK50A+cast iron	0.34—0.38	0.77—0.80	20	4

The use of cast irons as a counter body is associated with their high casting properties and, consequently, the simplicity and low cost of manufacturing elements of a friction pair, good machinability, high strength, no lead with thermal shocks and satisfactory durability. The use of steels as a counter body is explained by their high mechanical and thermal properties.

It should be noted that the task of selecting a more effective counter body for heavily loaded friction joints (for example, in the brakes of the aircraft landing gear wheels) is very relevant from the point of view of resource and safety. One solution to this problem is an approach in which both elements of a friction pair are manufactured by powder metallurgy methods, which allow varying the mechanical, thermal and frictional properties of the material to a wide range and obtaining, for specific operating conditions, the performance characteristics of the friction pair as a whole. For the first time, such an approach was implemented in England, where it was proposed to apply a sintered composition to the counter—body disks. In particular, the Dunlop's aeronautic brakes in the Concorde supersonic passenger aircraft have both tribopair elements made of sintered materials. Such brakes significantly exceed the smoothness of activation and lifetime of brakes containing alloyed steels as a counter body.

To work in conjunction with metal-ceramics FMK11, FMK79 and MVK50A, a friction material FMK845 (Fe-5C-2.2Si-1.2Ni-0.6Cr-0.8W-1Mn) with a hardness of 650 HRF—1 000HRF was developed. Tribological properties of such pairs are listed in Table 7.8.

TABLE 7.8 Friction properties of friction pairs from powder materials

Material brand	μ	Stability (Δf)	Wear during 1 braking/μm	
			Metal ceramics	FMK845
FMK11	0.241—0.309	0.72	22	6
MVK50A	0.330—0.404	0.70	16	12
FMK79	0.30—0.40	0.65	16	12

Iron-based friction materials are made by powder metallurgy by pressing at $P=500$—800 MPa, sintering under pressure $P=1.5$—3 MPa in H_2 medium at temperature $T=950$—1 050 ℃.

7.3.3 Cermets for Aircraft Brakes

In highly loaded chassis wheel brakes, cermet-based friction materials are used. The use of such materials in addition to the perception of large dynamic loads reduces the weight and dimensions of the structure, which is important in aircraft construction[35].

The binder component of such a composite is most often copper alloys with good thermal conductivity, and in special cases, iron and its alloys. Borides, carbides, nitrides, oxides and their mixtures are used as ceramic phases. The content of the ceramic phase usually exceeds 40%. Combining plastic metals with durable but fragile refractory compounds in a cermet is possible to obtain materials with better properties than the initial components[119].

Friction materials based on cermets are quite heat-resistant and operate at a temperature of about 1 000 ℃. The coefficient of friction, depending on the operating conditions and the applied counter body, varies between 0.3 and 0.7.

Such materials are characterized by high wear resistance and heat resistance[129].

The high content of non-metals in the cermet leads to a decrease in their strength. To improve the strength characteristics of cermets, a special technology is needed, consisting of double pressing and sintering, hot pressing and sintering under pressure.

Highly alloyed heat-resistant steels are used as the material of the counter body to work with friction cermets, or a wear-resistant layer with special properties is created on the surface of the counter body. For the cermets, sintered, heat-resistant composites, similar in composition to the material of the friction element, are also selected as the second element of the friction pair. The advantages of such a pair of friction are to equalize the tribomechanical characteristics and improve workability, which allows to transfer a higher braking torque and increase the durability of friction parts.

For manufacturing multi-disc aircraft brakes, an effective cermet has been created with composition Fe-23Cu-6.5Mn-2.5MoS_2-6.5BN-10B_4C-3.5SiC. This material has the following physical and mechanical properties: $\rho = 5\,900$ kg/m^3, $R_m = 40$—50 MPa, $R_{m,com} = 150$—250 MPa, $\tau_{sh} = 80$—100 MPa, $\alpha = (11.0$—$12.3) \times 10^{-6}$/℃ (in the temperature range $T = 20$—500 ℃). The frictional characteristics of this material when working in a pair with cast iron are presented in Table 7.9.

TABLE 7.9 Friction properties of Fe-23Cu-6.5Mn-2.5MoS_2-6.5BN-10B_4C-3.5SiC cermet paired with cast iron

Specific braking power/ [MPa · (m · s)$^{-1}$]	P/MPa	μ	Stability(Δf)	Wear during 1 braking/μm	
				Cermet	Cast iron
1	0.16	0.52	0.88	2.3	1.8
2	0.54	0.31	0.82	4.7	4.7
3	0.95	0.26	0.75	2.0	4.7
4	1.33	0.22	0.69	6.0	8.2

In China, the copper-based cermet with a steel substrate is patented and successfully being used in aviation brakes. It has the following composition: Cu-(10—24)Fe-(9.5—12.0)Sn-(0.6—2.5)Mn-(1—3)MoS_2-(10—12)C-(5—10)SiO_2-(1—3)SiC-(1—3)Al_2O_3.

7.3.4 Composite Materials with Carbon and Ceramic Matrix

For operation in super-heavy conditions in the aviation industry, friction materials from *Carbon-Carbon Composites Materials* (*CCCM*) and *Ceramics*

Composites Materials (CCM) are used. These are friction materials reinforced with carbon and other fibers and belong to the latest generation materials. The benefits of these materials are:

① a relatively low density of materials, which allows reducing the mass of braking systems by 40%—60%;

② high wear resistance, which allows reducing the number of replacements and repairs;

③ the ability to absorb a large amount of kinetic energy by converting it into heat during braking;

④ high resistance to thermal shock (almost no sensitivity);

⑤ high permissible operating temperature (above 1 000 ℃).

The technological scheme for the production of carbon-carbon composites and ceramics composites materials includes the operations of obtaining a fibrous skeleton (preform) with a random or, more often, an oriented arrangement of fibers; saturation (compaction) of the framework with a matrix material; machining the resulting workpiece (if necessary).

Saturation of the skeleton with a matrix material, which is the most crucial operation, is carried out with various solid, liquid and gas-phase operations, or a combination of them.

Solid-phase methods, which belong to powder metallurgy, prepare a mixture of the matrix material, reinforcing filler in the form of short fibers, whiskers or carbon nanotubes and a small amount of polymer binder. The mixture is pressed and heat-treated or HIP is performed. These methods are used mainly for the production of composites with a ceramic matrix and the technology is a modification of the powder method for producing technical ceramics.

Liquid and gas-phase methods are based on compaction of porous fiber cages heated to certain temperatures during filtration through the liquid (for example, Liquid Silicon Infiltration, LSI) or gaseous (for example, Chemical Vapor Infiltration, CVI) chemical precursors through them. In the preparation of the carbon matrix, gaseous hydrocarbons (methane, propane, ethylene, benzene, acetylene, etc.) are used as chemical reagents (precursors). In the production of composite materials with a ceramic matrix using CVI methods, the possibility of using various ceramic materials as a matrix is being researched. Composite materials with a SiC matrix have been widely used.

While producing composite materials, five modifications of the CVI method were developed: isothermal, thermogradient, isothermal, thermogradient with forced filtration and pressure pulsation methods. These modifications differ in the mechanism of the transfer of gaseous components in the porous frame. The most practical application was found by the simplest options—isothermal and thermogradient CVI methods.

The isothermal method is implemented in a uniformly heated chamber. The formation of high-quality composite material takes a long time (up to several months). The thermogradient method is characterized by the presence of a temperature gradient over the thickness of the framework. This method allows us to obtain a composite material with a high density per cycle and is characterized by a relatively high deposition rate and the possibility of volumetric compaction of the material.

There are two options for the implementation of the process of liquid-phase compaction of the fibrous frame. The first options is based on the processes of high-temperature pyrolysis of the corresponding organometallic precursors based on thermoplastic and thermosetting polymers, which impregnate the porous fibrous framework. To obtain a carbon matrix by this method, thermosetting phenolic and furfuryl resins and coal tar or oil pitches are usually used as starting materials. Polycarbosilanes or other Si-containing polymers are used for production with a SiC matrix. As a result of the subsequent heat treatment, polycarbosilanes decompose with the release of a solid residue enriched in SiC. The second option of the liquid phase method is used to obtain a composite material with a SiC matrix. The method is based on the mechanism of the reaction sintering of the workpiece material in the process of siliconization—impregnation with molten silicon (LSI).

As applied to the fiber-reinforced ceramic composite materials, the liquid-phase technology has several modifications differing in the method of introducing carbon filler into the fibrous framework. The most widespread modification technologies are:
- impregnation of the reinforcing material with a suspension containing carbon-graphite powder;
- impregnation with a polymeric binder followed by carbonization of the binder and the formation of a carbon matrix in the interfiber space of a coke.

The second modification is considered as the most promising for obtaining friction ceramic composite materials with a SiC-matrix. The use of the technology of pressing blanks, based on the use of short fibers, allows to significantly reduce the cost and simplify the production of friction parts. In addition, quasi-isotropic composites with short fibers have a higher thermal conductivity in the direction perpendicular to the friction surface, compared with layered structures based on fabrics or unidirectional tapes. This affects the decrease in temperature in the contact zone during braking, which leads to an increase and stabilization of the coefficient of friction, as well as a decrease in wear.

To improve the performance properties of brake devices, researchers are particularly attracted by gradient friction composite materials, which combine high wear resistance of surface layers with high internal toughness.

Ceramic-composite materials are used to increase the braking efficiency when the temperature on the friction surface can reach 1 000—1 200 ℃. The main disadvantage of materials based on ceramics is the low resistance to thermal and mechanical shock loads, which limits their scope. In friction joints, friction composites with carbon reinforcing filler (chopped fibers, fabrics, tapes) and a SiC-based ceramic matrix, obtained by impregnating a carbon blank with molten silicon, have been found the greatest use. Such friction materials have good tribomechanical characteristics—a high and stable coefficient of friction, weakly dependent on the speed of movement ($\mu=0.6$—0.7 at the beginning of deceleration decreases to $\mu=0.45$—0.55 at a speed of $v=15$ m/s), friction surface temperature (decreases by 5%—15% at a temperature of 500 ℃), contact pressure (decreases by 10%—20% with an increase in pressure up to 4 MPa) and weather conditions.

By design, brakes made of CCCM come with a steel bearing base (similar to metal powder) or without a steel base (so-called mono-discs), which are preferred in operation.

CCCM is used in high-loaded friction joints for which high heat generation is typical, and has a high friction coefficient ($\mu = 0.4$), especially at elevated temperatures. CCCM is characterized by low sensitivity to the specific contact pressure on the surface of the friction pair, speed, humidity, and it is quiet while working. In addition, there is a lack of setting of brake discs for the CCCM. The main disadvantage of CCCM as a friction material is its low coefficient of friction ($\mu=0.05$—0.28) at negative and low (up to 100 ℃) temperatures. In this connection, either different step of braking modes, or heating of the brake material by several preliminary braking are used. To increase the initial friction coefficient abrasive powders (SiC, Al_2O_3, B_4C and others) can be added to the polymer binder, from which a carbon matrix is obtained after pyrolysis, or the fibers themselves can be modified by applying ceramic coatings on their surface using various methods. In this case, the initial coefficient of friction can be increased by more than two times, but there is a decrease in the working coefficient of friction. Another disadvantage of these materials is the increased oxidizability of carbon fiber and the matrix at temperatures over 500 ℃ and the presence of oxygen, which leads to accelerated degradation of the friction surface and a decrease in the average wear resistance at different temperature and power conditions of the braking systems.

However, carbon matrix materials are widely used in the production of braking devices in aeronautical engineering. Currently, more than 60% of all operating aircraft landing gear are equipped with CCCM system brakes. While the relative need of the CCCM in rocket and space technology is constantly decreasing, the volume of production of brake discs for aviation is growing by ~12% per year.

In addition to aviation, carbon-carbon brakes are used in space shuttles, racing cars, high-speed trains, and other devices that require extra working conditions. In Formula 1 sports cars, only carbon-carbon brakes can withstand the stresses caused by extreme braking.

Chapter 8 Special-purpose Functional Materials

8.1 Concept of Functional Materials, Their Role in Engineering

The concept of *functional material* implies that the specific material has, in addition to the mandatory list of mechanical properties, additional special characteristics that are not typical for standard structural materials. The term "functional materials" refers to a wide class of substances that are used in the most diverse areas of modern life: from microelectronics and medicine to space research, and have quite specific, preferably adjustable, physical and chemical properties. Composites, alloys, ceramics, polymeric compounds can be referred to such materials. Creating one optimal material requires enhancing or suppressing some properties in existing material, that is, researchers deal with the design of a material to meet specific service requirements. Thanks to functional materials, it is possible to realize the performance characteristics of structures and devices that are not achievable when using other materials. An example is Nitinol in medicine. This is a unique combination of special (functional) properties of shape memory with high corrosion resistance in human body fluids and features of super elastic mechanical behavior similar to the mechanical behavior of bone tissue. This ensures complete biocompatibility and makes Nitinol indispensable for special implants, among other materials, which have a shape memory effect.

The most well-known groups of functional materials include composites with special electrical, magnetic, and optical properties. A typical example of composites, for which electrical properties are of decisive importance, is dispersoids, in which the ionic transport almost completely flows along the interphase boundaries of the matrix and the filler.

Functional nanocomposites also include many of the modern high-temperature superconducting materials, in which an increase in the critical current is achieved by introducing nano-inclusions of other phases thereto in order to create additional pinning centers. The implementation of the effects of giant and tunnel magnetoresistance and devices based thereon is possible only with the presence of multiphase structures—magnetic nanocomposites.

Compared with other fields of science and technology, new and modern functional materials play a very specific role, which can be characterized as follows:

① Materials are the core of the implementation of scientific and technical ideas, determining the possibility of such an implementation;

② Materials are at the forefront of scientific and technical progress in general, as they contribute to the realization of new ideas when the need arises from materials with new characteristics;

③ The competitiveness of industrial products in the era of high technology is determined by the level of achieved characteristics of the materials in use.

For the application in aerospace engineering, the most important functional materials with special properties, which are produced by powder metallurgy techniques, are heavy alloys, porous materials, sealing materials, radio transparent materials, optical ceramics and materials with special electromagnetic properties.

8.2 Heavy Alloys

Heavy alloys include materials with a density $\rho \geqslant 16\ 500\ kg/m^3$. The density of heavy alloys depends on their chemical composition, which, as a rule, includes up to 90% of tungsten. The alloying phases in heavy alloys are the Ni-Cu, Ni-Fe and Ni-Cu-Fe components. The chemical composition and properties of the most common heavy alloys are given in Table 8.1[118].

TABLE 8.1 Composition and physical and mechanical properties of heavy alloys

Chemical composition				Properties		
W	Ni	Cu	Fe	$\rho/(kg \cdot m^{-3})$	R_m/MPa	$A/\%$
91.8—92.8	4.5—5.1	2.7—3.1	—	17 000	—	—
89.2—90.2	6.8—7.4	—	3.0—3.4	16 600—17 100	700	3
88—92.3	5.0—7.0	—	2.7—5.0	17 000	800	3
90	10			17 000	826	7
97.4	2.6			18 500	770	4—8

The most widespread heavy alloys are the ones with Ni-Cu(Ni:Cu ratio of 2:1) and Ni-Fe with Ni:Fe ratio of 7:3—1:1[118]. A number of heavy alloys containing Fe, Cu, Mo, Ni are registered under the trademarks "Anviloy" and "Giromet".

Since tungsten has the highest melting point among the elements discovered, it is practically impossible to produce tungsten alloys by the melting method; therefore, heavy alloys are made exclusively by powder metallurgy methods. At the same time, people strive for 100% filling of the specimen volume, which is achieved either by impregnation of a porous tungsten specimen or by sintering with a liquid phase. The minimum initial porosity for the impregnation of the workpiece is 10%, while after impregnation the residual porosity is within 1%. The best option for obtaining high final density is sintering with the liquid phase when the mixture is prepared not by mechanical mixing, but by the chemical precipitation of the components. In this case, tungstic anhydride is impregnated with nitrates or sulphates of copper, nickel and iron, subsequently decomposed in air and reduced in hydrogen. In this way, a fine and homogeneous distribution of the precipitates is obtained. In addition, the powders are alloyed to improve their sintering.

The manufacture of a heavy alloy with a composition of W-Ni-Cu is carried out according to the following technology. The powders are compacted with the addition of 1.5% glycerol (to prevent delamination) on hydraulic or mechanical presses. More promising is Hot Isostatic Pressing (HIP) process, providing high-density briquettes. Sintered in two stages. At the first stage ($T = 1\ 000 - 1\ 100\ ℃$), with a slow rise in temperature, the removal of glycerol, adsorbed gases and the diffusion of W atoms into Ni occurs with the formation of the second phase. At the low-temperature sintering stage, the processes of spheroidization and coalescence of the pores happen. The second high-temperature stage of sintering is carried out at a temperature of $T = 1\ 480 - 1\ 500\ ℃$ in the environment of hydrogen. An increase in temperature causes the appearance of a liquid phase, which intensifies the sintering process. At this stage, the structure of the alloy is formed, which determines its physical and mechanical properties. The structural formation of the alloy is associated with the processes of W diffusion in Ni, a quantitative increase in the second phase, compaction, and recrystallization. The characteristic equiaxed grains of tungsten indicate the occurrence of a recrystallization process in the presence of a liquid phase.

A variation of the W-Ni-Cu alloy of high density is an alloy with an addition of 0.3% Pb or Bi. These small additives improve the machinability of the alloy without affecting the manufacturing technology and the physical and mechanical properties of the alloy[7].

High-density alloy system W-Ni-Fe is made as follows. After sufficiently mixing, the powder mixture of Fe-Ni with a weight ratio of 23:17 to 3:7 is added to the matrix metal-containing more than 90% of W. The mixture after mixing is pressed at a pressure of $P = 200$ MPa and sintered in hydrogen at a temperature of $T = 1\ 480 - 1\ 520\ ℃$. The microstructure of the alloy W-Ni-Fe consists of grains W and solute W in Ni and Fe. The grains have quiaxed shape characteristic of recrystallization in the presence of a liquid phase. High ductility of the alloy is provided by an interlayer of solid solutions formed around the tungsten grains. Ni-Fe heavy alloys have higher mechanical properties than Ni-Cu alloys. The addition of Co to the alloys with a ratio of Ni:Co:Fe=5:3:2 increases the hardness and strength but reduces the ductility of the material[35].

Unprotected heavy alloys do not oxidize up to a temperature of 450 ℃. After applying a special chrome coating, the alloys are stable in the air up to 930 ℃. The heat resistance of such alloys is quite high — they withstand hardening from a temperature of 1 040 ℃ without cracking. Heavy alloys are well machined. The main task in the manufacture of heavy alloys is to achieve minimum porosity to ensure qualified density and mechanical characteristics.

The use of heavy alloys in aviation and rocket and space technology is associated, first of all, with their high density—they are used where the concentration of large mass in a minimum volume is necessary. Heavy alloys are used in counterweights of gyrocompasses, as anti-blatter cargoes for ailerons of airplanes, for the manufacture of parts of balancers and flywheel. They are also used in propeller blades, radar equipment and aviation electronic systems. Heavy alloys are sold under the trademarks "DENSIMET" (W-Ni-Fe) and "INERMET" (W-Ni-Cu).

When choosing a material for a particular application, one should take into account that alloys with iron are more durable, but they are ferrimagnetic, and alloys with copper do not have magnetic properties and therefore they are preferred for the manufacture of gyro wheels.

8.3 Porous Permeable Materials

Porous materials are materials with a characteristic feature of purposefully organized porosity. Depending on the purpose, the porosity p in such materials can be regulated within wide limits. Conventionally, these materials are divided into:
① low porous ($p \leqslant 30\%$);
② medium porous ($p=30\%-50\%$);
③ highly porous ($p > 50\%$).

Porous materials are widely used in various fields of technology, in particular aerospace. As components that require high permeability, materials with medium and high porosity, with a greater proportion of open porosity are used. Some, the most characteristic areas of use of porous materials in aerospace engineering, consider below.

8.3.1 Filters

In aeronautical engineering, porous permeable materials are most often used for manufacturing filter operating elements. To obtain metal filters, the initial components in the form of powders or fibers are used. The advantages of metal filters are a wide range of porosity and permeability, sufficiently high strength, insensitivity to shock loads, heat resistance, corrosion resistance in aggressive media, good thermal conductivity, and the possibility of regeneration. Porous powder and fiber metallic materials from which filters are made can be subjected to various types of machining, welding and brazing.

Metal filters are equipped with many vital systems of the aircraft, as they ensure the reliability of all units and flight safety. The examples of the filtering elements in various purposes are given in Fig. 8.1.

FIGURE 8.1 Filter elements made from powders and fibers for various purposes

The main systems of the aircraft in which filters are used include:

① anti-icing system and air conditioning system, filtration of hot air drawn from the engine and supplied to the front edges of the wing and tail assembly, as well as into the cockpit and passenger compartment after adjusting to the set temperature. Filtration is performed to prevent clogging of the holes in the microinjector tubes and to ensure the safety of passengers and crew.

② fuel system, for filtering fuel to prevent the failure of fuel pumps and engine injectors.

③ hydraulic system for filtering hydraulic fluids and preventing the failure of parts (spools, valves, hydraulic cylinders, chokes, etc.). The hydraulic system, along with the fuel, is one of the most critical systems of the aircraft.

In connection with known cases of accidents due to jamming of the actuators of fuel and hydraulic systems, the purity of the fuels and other working fluids is very strictly controlled. A whole system of control and cleaning has been developed during the passage of fuel from the rail tank car to the aircraft's tank (receiving tank from the rail tank car, storage, service tanks, oil sealed transformer, aircraft tank). At each stage of transfer of fuel from one tank to another, filtration and control of the purity of the fluid are performed. In case of draining the fuel from the caisson of the aircraft, it returns to the receiving tank. Sometimes the use of fuel drained from an aircraft is not allowed and it is recycled. In this regard, special attention is paid to aviation filters and materials for their manufacture.

The filter materials must have the following basic properties:

① uniform porosity and pore size corresponding the fineness of cleaning, to ensure the specified filtration quality;

② minimal hydraulic resistance;

③ maximum "dirt holding capacity";

④ possibility of regeneration;

⑤ resistance to mechanical stress;

⑥ corrosion resistance, including in aggressive environments.

Since no filter material can satisfy all the requirements at once, the filter design usually consists of several filter elements and the cleaning proceeds are conducted by stages (that is, cascade filters are used). The highest demands are referred to the filter element at the outlet of the filter—these are fine filters. The fineness of cleaning (the maximum permissible particle size of the pollutant) of aviation filters is 5 μm and they should work at temperatures up to $T = 200—300$ ℃. Metal gauzes, manufactured by the industry, provide cleaning accuracy of 10—20 μm. Paper filters of special types, which provide the necessary filtration but do not have sufficient strength, have a greater hydraulic resistance and quickly fail at temperatures above 100—120 ℃. In aviation, fine filter elements are manufactured by powder metallurgy from porous nickels or stainless steels.

In the production of aviation filters, two types of metal powders are used—crushed (non-spherical) and sprayed (with a spherical shape of particles).

The *benefits* of spherical powders are:

① low resistance;

② high penetration;

③ easy regeneration.

The *disadvantages* of spherical powders are:

① low strength;

② the degree of purification is less than for non-spherical powders.

Non-spherical powders are obtained by rolling and sintering powders into a tape, followed by crushing the tape with special mills and separation into fractions.

Filters made from metal fibers have the best properties, and most often they are stainless steel fibers. Such filters have high permeability, uniform porosity, high cleaning fineness, which depends on the fiber diameter, and high mechanical strength compared to powder. The technology of their manufacture is similar to powder filters. The technological scheme for obtaining nickel filter elements is given in Fig. 8.2.

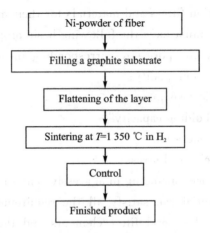

FIGURE 8.2 Manufacturing technology of filter elements

Nickel powder filter elements have all the necessary service characteristics, but insufficient strength. The most popular porous material for manufacturing aviation fine filters is rolled steel of corrosion-resistant steel made from sintered powders of the FNS-5 and FNS-2-3 brands, which have a cleaning accuracy of 5—6 μm and 2—3 μm, respectively.

For manufacturing this material, the specialists use powders with size of 0.04—0.06 mm for the FNS-5 and 0.02—0.03 mm for the FNS-2-3, obtained by the reduction of oxides with calcium hydride. The technological scheme of production of filter elements made from stainless steel Fe-18Cr-15Ni is given in Fig. 8.3.

Rolling of powders in a porous ribbon is carried out on a two-roll vertical rolling mill. After sintering the ribbon, filter material from austenitic steel (Fe-18Cr-15Ni) with 0.14—0.20 mm thickness is obtained, which ensures high

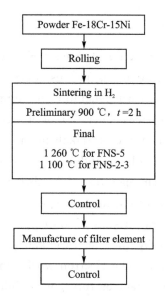

FIGURE 8.3 Technological scheme for manufacturing rolled powder sintered steels (grades FNS) and filter elements therefrom

filtration efficiency of working media and, accordingly, reliability and long-term trouble-free operation of vital aircraft units. However, they are relatively easy to regenerate and allow multiple ultrasonic washing. Some key characteristics of the austenitic steels filter elements are as follows:

① Porosity:
- FNS-5 is 36%—42%;
- FNS-2-3 is 22%—30%.

② Carrying capacity:
- FNS-5 for aviation fuel TC1 is 0.017—0.023 1/(min·cm^2);
- FNS-2-3 for aviation fuel TC1 is 0.003—0.006 1/(min·cm^2);
- Working temperature range is $T = -60$—250 ℃.

Promising for use in aerospace are filters obtained from a mixture of powders of titanium hydride (TiH_2) and aluminum acetic acid $Al(C_5H_7O_2)_3$. Powders in a ratio of 3:1 are thoroughly mixed, pressed under pressure $P = 300$—500 MPa and sintered at $T = 1\ 350$ ℃ in vacuum for 3—4 h. After sintering, blanks with uniformly distributed porosity of ~50% are obtained. The pore size depends on the granularity of the starting titanium hydride powder. The strength of the material R_m is ~100 MPa. The material is well processed by cutting. Permissible operating temperature is up to 500 ℃. The material allows repeated regeneration, and can work in harsh environments[131].

8.3.2 Porous Ionizer

Due to the progress in rocket and space technology, porous materials have found another popular application—as porous ionizers for ion-powered rockets. Such engines can be used as thrust generators for long-range spacecraft. In ionic engines, a nuclear reactor provides generation of electricity consumed for ionization and acceleration of propellant particles—alkali metal vapor, which creates thrust. Ion engines make it possible to obtain high flow rates of the working fluid with low fuel consumption and long-term operation. The thrust of such engines is small, so they can only be used for movement in outer space.

All the performance characteristics of ion engines (efficiency, traction force, service life) are determined by the electrical and physical properties of porous ionizers that generate ion flow. The electron work function, ion current density, neutral component, ionization temperature and stability of the porous structure during prolonged high-temperature operation are the main criteria for the quality of porous ionizers. Porous tungsten alloys with the smallest possible pore sizes and maximum concentration most fully meet these requirements. The use of porous tungsten (including doped with rhenium) made of micro- or nano-powders provides almost complete ionization of alkali metal vapors.

The most effective way to compact tungsten powder is a semi-isostatic pressing method—in elastic (rubber) shells, using steel molds. The compressibility and formability of tungsten powder are improved by annealing in hydrogen at a temperature of $T = 1\ 150—1\ 200\ ℃$ or by the introduction of surface-active substances that do not contain carbon compounds. More preferred is the first method. Billets are sintered in vacuum at a residual pressure of $P = 10^{-3}—10^{-4}$ Pa. The furnace is slowly heated to a temperature of $T = 800\ ℃$ in order to remove volatile components. Subsequently the heating rate is increased to the maximum possible for the furnace. The preferred temperature of isothermal heating is $T = 1\ 800—2\ 100\ ℃$[7]. The highest compaction intensity is observed during the first hour of heating. The increase in the sintering time accompanying with a much lesser extent affects the compaction of products, and after 4 h the compaction process practically stops. When choosing the sintering mode, one should be guided by the given optimal density and geometrical parameters of the pores (shape and size) of the future material, on which the characteristics of the ionizers produced from this material depend.

The performance of the ionizer during its long-term operation under conditions of high temperatures (up to 1 200 ℃) is significantly affected by the stability of the organized structure of the pores. The criterion for the stability of the porous structure is the amount of permeability loss over the design time of operation (10 000 h), equal to 50% of the initial permeability. The decrease in the stability of the ionizer is a consequence of change in its porous structure.

The output of the neutral component at identical ion current densities for ionizers with an unorganized porous structure is an order of magnitude larger than that for ionizers with an organized porous structure. In addition to the nature of the porous structure, the output of the neutral component, the electron work function and other characteristics of the ionizer are also significantly influenced by the microstructure of the sintered tungsten powder and the content of impurities, especially carbon, in the form of carbides WC and W_2C. The necessity of cleaning tungsten powder from carbon follows from a comparison of the emission characteristics of tungsten and its carbides: for tungsten $\varphi=4.55$ eV, for W_2C $\varphi=3.85-4.80$ eV, for WC $\varphi=2.24-3.60$ eV.

The decarburization of tungsten powder can be achieved by heating in an environment of oxygen or water vapor, as well as by evaporation at temperatures above 2 450 ℃. Doping tungsten with rhenium also helps to improve the performance of ionizers[134].

While manufacturing tungsten ionizers, it is necessary to take into account the presence of impurities that affect the work function of electrons, in particular Ni, Al, C, and to prevent them from entering the material in every possible way when performing technological pressing and sintering operations[134].

8.3.3 Transpiration Cooling Materials

Transpiration cooling materials are porous materials that are used to cool high-temperature components of aircraft. The cooling of the working surfaces of products made of such materials occurs as a result of transpiration of liquid or gases through the pores. In general, there are two categories: evaporative-transpiration and self-transpiration cooling materials[135].

Evaporative-transpiration cooling materials are materials where active thermal protection is provided by forcing the liquid or gaseous refrigerants through porous or perforated materials. Gaseous refrigerants move towards the heat flow and cool the material due to heat absorption by the gas. In the case of liquid refrigerants, the effectiveness of cooling is improved since heat absorption occurs due to evaporation or dissociation of the liquid. In the case of the evaporative-transpiration cooling method, the main factor for active thermal protection is the so-called thermal blowing effect. It lies in the fact that refrigerant in the gaseous state drives back from the working surface, dramatically changing the temperature and velocity profiles in the boundary layer and thereby reduces the heat transfer coefficient, which leads to a decrease in heat transfer rate.

The main characteristics that determine the performance of evaporative-transpiration fluids are:

① permeability, determined by the open porosity, the geometry of the through channels and the degree of roughness of their surface (evaporative-transpiration cooling materials are porous materials with prevailing open porosity);

② heat resistance (increase in heat resistance reduces the consumption of refrigerants due to the increase in permissible temperature and, consequently, the efficiency of cooling);

③ uniform permeability over the entire working surface of the material (homogeneity of local permeability).

The evaporative-transpiration cooling materials are made by powder metallurgy from powders or fibers. The advantages of fibrous materials compared to powder-sintered materials are:

① a significantly larger proportion of through porosity (fewer closed and dead-end pores) with an equal total porosity;

② higher strength and heat resistance with equal porosity;

③ a smoother surface of the through channels (determined by the initial surface roughness of the fibers) and, as a result, a decrease in hydraulic resistance;

④ a higher degree of homogeneity of local permeability due to the presence of oriented ("organized") structure;

⑤ significantly higher impact strength.

For use in moderately harsh operating conditions for parts of the complex spatial shape, it is preferable to use powder transpiration cooling materials, and in especially harsh operating conditions—fibrous materials.

Fibrous materials are made of relatively short chopped fibers of 2—7 mm long or continuous fibers, whose semi-finished products are woven or knitted nets. When using short fibers, the material is molded by the method of liquid felting. It consists of preparing a suspension of fibers in a liquid (suspension), followed by filling the appropriate form with a suspension. Usually under a permeable base on which fibers are deposited, a vacuum is created, or the capillary forces of the base material (mold) are used to remove the liquid by slip casting. It is also possible to vibrate the pre-wetted fibers for uniform compaction with a volatile liquid. When using meshes, the processes include laying or winding them in the correct orientation. Further processing involves the use of traditional methods of powder metallurgy—cold pressing or rolling followed by sintering, sintering without pre-pressure treatment, pressure treatment after sintering, etc.

Porous cooling, especially with the use of liquids, is promising in aircraft engines and during aerodynamic heating of various aircraft. One of the most commonly used materials for this purpose is Nichrome (25Cr-75Ni) with a porosity of 40%, which is designed to work in the parts of aircraft subjected to high temperatures, to partially remove heat flux from a cooler by penetrating through pores. This material is effective for cooling by liquids or gases transpiration of thermally loaded nodes (combustion chamber walls, intra-complex compound nozzles), operating at temperatures up to 1 000 ℃, and also works as facing material in anti-icing systems. The physical and mechanical properties of porous Nichrome made by powder metallurgy technology are given in Table 8.2.

TABLE 8.2 Physical and mechanical properties of porous Nichrome 25Cr-75Ni at different temperatures

$T/℃$	E/GPa	R_m/MPa	$A/\%$	$KCU/(\text{kJ}\cdot\text{m}^{-2})$	HB/MPa	$\rho/(\text{kg}\cdot\text{m}^{-3})$
20	33.5	206	1.2—1.25	55	125	4 800
700	—	101	1.2—1.75	—	—	—
800	—	62	—	—	—	—
900	—	59	—	—	—	—
1 000	—	13	2.0—3.0	—	—	—

Self-transpiration cooling materials are one category of transpiration materials, which are pseudo-alloys, consisting of two components—a refractory frame, the pores which are filled with a high-enthalpy filler that does not interact with the frame material. In such materials, the main heat-absorbing effect is due to phase or chemical transformations of the filler. Low-melting metals with a melting or boiling point below the melting temperature of the framework material are used as a filler, as well as chemical compounds that sublimate or decompose with a significant endothermic effect. In addition to heat absorption by phase transitions, the effective thermal protection with the use of self-transpiration cooling materials also results from:

① high thermal conductivity of the frame material, providing intensive heat removal from the working surface and dissipation of the heat from contacting parts in a large mass;

② high heat capacity of the material, resulting in the heat accumulation primarily in the material itself;

③ blowing effect in case of availability of gas and vapor products from phase or chemical transformations.

The advantages of self-transpiration cooling materials over evaporative-transpiration cooling materials are that there is no need to use a storage tank for the refrigerant and no necessity of supplying it to the work area. The disadvantage is the strictly limited time for their operation, determined by the volume of the ablating component and the evaporation rate from the pores of the frame under specified operating conditions.

Self-transpiration cooling materials are manufactured by powder metallurgy methods using two technologies: preforming the refractory framework by pressing and sintering, or by hot pressing, followed by impregnation with active components; direct hot pressing of the compositions, provided that the thermodynamic characteristics of the active ingredient permit.

The service of rocket nozzles is accompanied by loss of their original size and shape due to oxidation, erosion, thermal shock and a complex stress state, which places high requirements on the materials for nozzles. For the manufacture of nozzles using tungsten, its filler with high enthalpy not only absorbs heat due to endothermic transformations but also reduces the heat load as a result of the formation of a steam jacket. Those functions ensure the maintenance of the gas working temperature at the required level and dimensional stability in the nozzle throat[134].

In addition to the enthalpy properties of the filler, the performance of the nozzle is greatly influenced by the material of the porous framework—its strength, pore size and shape, uniformity and density. The optimal technology for the manufacture of porous tungsten blanks is HIP—it gives a stable reproducibility of the properties of the blanks on the density and uniformity of the pores. The density of the blanks is one of the important characteristics since it determines the self-cooling capacity of the rocket nozzle. The optimum density of the blanks corresponds to $0.75\rho_{th}$—$0.78\rho_{th}$. With such density, the formation of closed pores is practically eliminated, which dramatically reduces the erosion resistance of the material. In rocket nozzle blanks with a density of $0.72\rho_{th}$ and $0.78\rho_{th}$, the range of the pore sizes is the same (3—10 μm). The nonuniform distribution of the pore size in the material is observed depending on its density. The consequence of the difference in pore size includes the nonuniform permeability of the working surface of the product and the difficulty to control the repeatability of its structure and properties during manufacture.

The infiltration of tungsten blanks is carried out in the environment of hydrogen or argon. The infiltration temperature of tungsten blanks with an average density of $0.75\rho_{th}$ using filler Ag is $T=1\ 050$—$1\ 100$ ℃; for Cu $T=1\ 200$—

1 220 ℃; for Pb $T=325—330$ ℃; for Sn $T=700—725$ ℃ (H_2 medium with dew point not lower than -50 ℃). The infiltration time is established experimentally depending on the framework size and the physical properties of the infiltrate.

The degree of infiltration of tungsten blanks depends on the physicochemical properties of the infiltrate, which is chosen according to its enthalpy characteristics and chemical activity with respect to the matrix. Table 8.3 gives the thermophysical properties of the most used infiltrates[134].

TABLE 8.3 Thermophysical properties of infiltrates

Infiltrate	$\rho/(kg \cdot m^{-3})$	$T_{mel}/℃$	$\Delta H_{mel}/(kJ \cdot kg^{-1})$	$T_{ev}/℃$	$\Delta H_{ev}/(kJ \cdot kg^{-1})$
Ag	10 500	960	108.8	2 180	2 350
Cu	8 960	1 083	213.4	2 580	5 410
Mg	1 750	650	368.2	1 107	5 440
Zn	7 130	419	112.2	906	1 800
Sn	7 290	231.9	59	2 430	3 010
Pb	11 337	327.4	24.3	1 750	860

The value of ΔH_{ev} is an important characteristic to assess the self-cooling ability of an infiltrate, which is determined not only by its heat absorption capacity during the phase transition from a liquid to a gaseous state, but also by heat absorption per unit volume of the infiltrate[134].

The strength of sintered tungsten blanks depends on their density, the characteristics of the initial powders including their morphology. Billet, obtained from powders with spherical particle shape, has lower strength compared to that from standard tungsten powders. After infiltration, the strength of the workpieces increases and considerably exceeds the strength calculated on the basis of the ratio of the components making up the composite material. Such an anomalous increase in strength is associated with the hardening effect of the infiltrate due to chemical interaction with the tungsten framework. A similar phenomenon is known from the practice of making carbide tools. With an equivalent cooling effect by infiltration, nozzles with higher strength can maintain longer time with dimensional stability in the nozzle throat under high-temperature erosion wear. With increasing temperature, tungsten-based infiltrated materials become weaker.

Thermal erosion wear also depends on the physical properties of the infiltrate. With identical characteristics of tungsten skeletons, silver infiltration can significantly reduce erosive wear compared with copper infiltration. Improvement in the erosion resistance of infiltrated tungsten matrices is achieved by ensuring the optimal properties of the framework (density, strength, size, shape and uniformity of pores).

Nozzle throat inserts for solid propellant rocket engines, operating under conditions of high temperatures, high pressures and thermal shock loads, are made of porous tungsten infiltrated with Ag or Cu. The infiltrated alloy W-Ag is used for manufacturing nozzles of "Polaris" solid-fuel rocket at the operating temperature of ∼ 3 000 ℃. The dimensions of the nozzles in the critical section exceed 500 mm[134].

8.3.4 Capillary Structures of Heat Pipes

Capillary structures of heat pipes are capillary transport of liquid media. Heat pipes are heat conductors capable of transmitting super-powerful heat fluxes at high speeds and allow to reduce heat resistance by tens and hundreds of times. The thermal conductivity of heat pipes exceeds the thermal conductivity of copper several thousand times. Heat pipes are used in devices where it is necessary to bring in or remove a significant amount of heat very quickly. The main application of heat pipes in the aerospace industry is the cooling of electronic devices in spacecraft.

Heat pipes have such unique properties as high thermal conductivity, the ability to work for thousands of hours, a variety of possible design solutions, a wide temperature range of operation. This makes it possible to use heat pipes in aerospace engineering—for thermal stabilization of electronic equipment, high-speed discharge of excess heat, uniform temperature distribution on radiation surfaces; for equalizing temperatures on the leading edge of the wing, temperature control in cosmonaut costume working in open space, etc.

Initially, heat pipes were made by sintering a porous layer of copper or iron powders to the inner surface of a heat pipe. The disadvantage of this method is the impossibility of obtaining a capillary layer with high porosity. Heat pipes with a capillary layer of metal fibers have superior properties. In this case, the layer made from metal fibers is baked to the internal surface of the heat pipe in the environment of hydrogen or dissociated ammonia. Such a layer is a metal frame with porosity. This technology allows creating a capillary layer with predetermined structural and hydrodynamic properties that provide design characteristics and a high degree of reproducibility. The main parameters determining the properties of the capillary layer are the geometric and thermal properties of the fibers, and the geometry of the porous layer.

Cu, Ni, Al, W, Na, W, Ti, Nd-Zr, W-Re, Mo-Zr-Ni-C, corrosion-resistant steels, and Inconel alloys are used as candidate materials for the heat pipe shells. The capillary layer is made of metal grids (Cu, Ni, stainless steel, bronze) perforated sheets, particles or fibers and metal felts.

Heat carrying agents are substances that are chemically compatible with the material of the capillary-porous structure and body. Water, alcohols and ammonia are compatible with stainless steel, distained water, alcohols, ammonia, acetone with Al. Propane, butane, freon, acetone, a mixture of ammonia and benzene are compatible with Cu and Ni. At high temperatures, heat pipes made from austenitic stainless steels and Ni can work with Na and K heat-carrying agents, while Nb alloys can work with Li heat-carrying agents. A technology has been developed for obtaining capillary structures of contour heat pipes by the method of non-shrinkable sintering of composite powder materials. With a few centimeters diameters, the capillary structure should have a pore size of 1—5 μm with a porosity of 50%—65%, close contact with the inner surface of the evaporator system and contain a system of vapor-discharge channels inside. The capillary layer is designed to direct sinter to the heat pipe body and vapor-discharge channels can be formed inside the capillary layer. This design of the capillary structure allows increasing the limiting heat flux in the evaporator several times. However, when sintering the powder capillary structure of a large outer diameter directly in the evaporator housing, there exists a problem of ensuring reliable mechanical and thermal contact of the capillary structure with the housing, due to shrinkage of the powder during sintering. This problem is especially serious when sintering capillary structures with small ($\sim$1 μm) pores.

In this regard, a promising method is the liquid-phase sintering of the capillary structure formed directly on the inner surface of the pipe from a two-component mixture of fine powders (Cu-Al, Cu-Sn, Cu-Mg, Al-Zn, Ti-Al, Ni-Al, Ni-Cu and others). A feature of this process is that, under certain sintering conditions, the formation of a liquid phase is not accompanied by shrinkage, but by volume growth of the powder compact parts. This technology allows to obtain capillary structures that have good thermal and mechanical contact with the heat pipe body with a porosity of $\sim$60% and an average pore size of several microns. The main characteristics of the heat pipes are given in Table 8.4[136].

TABLE 8.4 Characteristics of nickel heat pipes

Outside diameter/mm	Length/mm	Evaporative capacity/W	Pore size of capillary structure/μm
5	50	30	1—5
8	60	100	1—5
22	150	1 500	2—10
78	300	6 000	2—10

8.3.5 Porous Electrodes for Fuel Cells

Porous electrodes are the main functional components of devices for direct conversion of chemical reactions into electrical energy. Porous materials have a large specific surface area, leading to their application for various kinds of catalytic processes to accelerate chemical reactions, including in fuel cells.

Fuel cells are galvanic cells in which the generation of electrical energy takes place due to electrochemical oxidation of fuel. The main advantage of fuel cells compared to currently used combustion-based technologies is high efficiency reaching 75% and growing with decreasing load, unlike heat and power plants and internal combustion engines. Fuel cells, in which nickel sintered electrodes are used, are applied in space technology.

The main task in the development of a fuel cell is to ensure the rapid flow of electrode processes at low voltages, which is partially solved by the use of electrodes made of materials with high catalytic activity in the reaction. The second task is to bring the reactant gases to the "electrode-electrolyte" boundary as fast as possible. To do this, the electrodes are made in the form of thin films (single-layer or multilayer) of porous materials. The gas pressure and pore sizes are tuned so that the "gas-liquid" boundary exists inside the pores.

For fuel cells operating at a temperature of more than 100 °C, electrodes are obtained by pressing or rolling carbonyl nickel powders with particles of arbitrary shape and porosity 40%—50% with a pore size of 1—3 μm after sintering. If the operating temperature is less than 100 °C, the surface activity of the powder may be higher. The surface activity is increased by coating the rolled porous strip from powder materials with nickel powders. Such a "double" structure with large surface activity is characterized by a high porosity of 60%—70% and ensures the stability of the gas-liquid phase boundary.

Plates with high surface activity are also obtained from a mixture of nickel and nickel-aluminum powders in a 2∶1 ratio. In this case, the surface activity of the sintered compact parts of the mixture powders is achieved by treating them in KOH or NaOH, during which aluminum selectively dissolves into solution. The surface activity can be enhanced by additions of NiB or precious metals (Ag, Pt)[105].

8.4 Sealant Materials

Sealant materials are materials used for manufacturing seals in jet engines that are installed between moving and stationary components. Sealant materials should have sufficient gas-tightness, deformability, corrosion resistance, heat resistance, erosion and wear resistance with low hardness, workability and low coefficient of friction. It is almost impossible to combine all the required properties in one material. Therefore sintered and composite materials, whose manufacturing technology allows providing a combination of high-performance properties, are used.

The operational and economic characteristics of engines, like gas turbine engines, are largely determined by the size of the gap between the rotor and the stator of the engine—the smaller the gap, the higher the Coefficient Of Performance (COP). On the other hand, a very small gap can lead to the destruction of the turbine blades due to its thermal expansion. A significant amount of fuel can be saved during the operation of aircraft engines with efficient gas channel seals.

In order to avoid contact between rotating and stationary parts of the turbine, radial and axial clearances are established between the rotor and the stator. The efficiency of gas turbines depends on the size of the relative radial clearances, both at the end of the rotor blades and gas seals between the steps of the impellers. The minimum radial clearance is 1.5%—3.0% of the height of the blade. In the case of a stage with a low degree of reaction, each percent of the relative gap causes a decrease in turbine efficiency by about 1%, and in the case of the reactive stage, by 2.5%—3.5%[137].

Nickel-based sintered sealant materials used in aircraft engines operate under conditions of high temperatures, high gas-dynamic pressures created by streams with sonic and supersonic speeds, causing incursion into the materials of the sealing rings. Under these conditions, the chemical composition of the surface layers of metals and alloys significantly changes, which determines the change in the physical and mechanical properties and significantly affects the performance, wear resistance and anti-friction of the friction pairs. The burning of graphite in nickel-graphite sealant materials leads to the occurrence of porosity, erosive wear, oxidation of nickel in the ring, the appearance of open microcracks under the influence of heat cycles and bending stresses. One of the reasons for the evolution of erosion destruction is embrittlement during the oxidation of nickel under the influence of high-pressure gas stream. This disadvantage is found at the temperatures of 600—650 ℃. Oxidation leads to sharp increase in the hardness of the inserts (from 30HB—45HB to 260HB), which adversely affects the workability and wear resistance of the ends of the working blades, as well as the impeller labyrinths.

To ensure reliable operation and increase the service life of high-temperature turbines, new sealant materials should have high heat resistance, low hardness (30HB—40HB), which should not change during operation. At the same time, it is necessary that during contact, the ridges, without their own wear, were easily produced in the groove inserts, which significantly reduces the gas flow and improves the efficiency of the turbine.

For the high-temperature path of the engine, sealant materials based on Cr20-Ni80 nichrome have been synthesized with the introduction of 6% BN in the solid lubricants, which successfully operate at temperatures up to 900 °C, possessing high heat resistance, and good anti-friction properties. Characteristics of sealant materials on nickel and nichrome base are given in Table 8.5[137].

TABLE 8.5 Characteristics of sealant materials

Characteristics	Material	
	Nickel base	Cr20-Ni80 base
HB	30—35	30—45
$R_{m,ben}$/MPa	≥70	≥100
Heat resistance at $T=900$ °C at 100 h/ (mg·cm^{-2})	25—30	8—10
Beginning temperature of bulk oxidation in air/°C	700—750	≥900
μ at $v=20$ m/s and surface temperature $T=800$—1 000 °C	0.11 (0.24 during long work)	0.12 (stable in time)

Hot bench tests of the material based on Cr20-Ni8 at the gas flow temperature 980 °C at the turbine inlet showed the stability of the structure and chemical composition of the material, which ensures the preservation of the initial physical and mechanical characteristics and the reliability of its operation.

The seals are made of the material based on Cr20-Ni80 by rolling a mixture of nichrome and boron nitride powders, followed by sintering in a protective environment.

Material with composition of Ni-3BN-2SiO$_2$ has been produced for operation at temperatures of 900—1 000 °C, having the following characteristics: $\rho = 6\,200$—$7\,000$ kg/m^3; $p = 13\%$—17%; $\alpha = 16.4 \times 10^{-6}$/°C; $\lambda = 31.8$ W/(m·°C) in the range $T = 20$—1 000 °C. The mechanical properties of the material at high temperatures are given in Table 8.6.

TABLE 8.6 Mechanical properties of the Ni-3BN-2SiO$_2$ material

$T/°C$	$R_{m,ben}$/MPa	$R_{m,com}$/MPa	R_m/MPa	τ_{sh}/MPa	$A/\%$	KCU/(kJ·m^{-2})	HB/MPa
20	120—150	500—600	50—70	100—120	0.3—0.4	10—18	250—450
600	—	—	30—45	40—50	0.3—0.5	—	—
700	35—45	—	—	—	—	14—23	—
800	—	—	16—27	15—25	—	16—26	—
900	—	—	—	—	0.8—1.0	—	—
1 000	25—30	—	—	—	2.0—3.0	—	—

Ni-3BN-2SiO$_2$ sealant material is produced by powder metallurgy: cold-pressing at pressure of $P=300—700$ MPa and sintering at temperature $T=1\ 150—1\ 200$ °C in protective medium (H$_2$, Ar, N$_2$, dissociated ammonia). To give the final dimensions, it is possible to use rolling for the sintered billet.

Currently, new types of sealant materials are developed. These are materials with adjustable gap and "metal felts". The first type is materials based on intermetallic compounds with a high melting point and the use of silicides of Mo, W, Ta, Ti and B. The compounds are selected so that at operating temperatures its gaps in the volume decrease to the minimum due to oxidation of the sealant materials under the action of the gas flow. When reducing the gap for every 0.076 mm, step efficiency increases by 1%. To control the radial clearances for large deformations of the stator, it is advisable to use ceramic ring inserts made of Si$_3$N$_4$, which practically do not change their size at working temperatures.

Sealant materials from "metal felts" have unique characteristics—the combination of high porosity with the characteristics of high strength and ductility, high resistance to oxidation, abrasion and erosion resistance, which are necessary to create a new generation of sealant materials. Possessing high porosity ($p \geqslant 80\%$), low hardness and high resistance to oxidation, the material has high penetration characteristics, which eliminate the wear of the blades while ensuring the necessary penetration depth and therefore guarantees high-performance characteristics of the products.

8.5 Powder Materials for Electrical and Radio Technical Purposes

Materials for electrical and radio technical purposes are a wide class of powder materials, which are used in various systems and devices of aviation and space technology due to specific physical properties. As a rule, these are not construction materials and do not require any level of mechanical properties, but they must possess a given set of special physical characteristics that are necessary to perform special functions. Such materials used in aerospace engineering include optical ceramics, radio-wave transparent materials, magnetic materials, and others.

8.5.1 Optical Ceramics

Optical ceramics is a kind of transparent ceramics that transmits, without distortion, electromagnetic radiation. If ceramics transmits infrared (thermal) rays, it is considered as optically transparent ceramics in the infrared range. The term "transparent ceramics" is conditional, since the transmission degree of waves with different lengths may be different.

Polycrystalline transparent ceramics were first obtained in the 60s of the 20th century, although it is known that even ancient Chinese porcelain was transparent in thin layers. The first technical material capable of transmitting light rays was aluminum oxide (Al_2O_3). Afterwards, transparent ceramics from various materials were obtained: oxides (Y_2O_3, MgO, CaO, Sr_2O_3, ZrO_2, TiO_2), fluorides, nitrides.

The centers of light scattering in ceramics are grain boundaries, residual pores inside and impurity phases including interfaces, birefringent structures and so on. Therefore, the grain boundaries in optical ceramics should be optically perfect, that is, free from impurities and have the narrowest possible range of disorder. In addition, to increase the transparency of ceramics, it is necessary to reduce the number and size of pores. For example, for transparent Al_2O_3 ceramics, the concentration of pores with a size of less than 100 nm should be below 0.1 vol.%. The density of optical ceramics should be close to theoretical value.

Optical ceramics exceeds the corresponding single crystals in their physical properties, functional and operational characteristics, and their manufacturing technology is much simpler and more economical than growing single crystals.

The traditional approach to creating optical ceramics is based on the HIP or vacuum sintering of micron (nanoscale) powders. Sintering is carried out at high temperatures (from T_{sin} to $0.7T_{mel}$), at which the processes of collective recrystallization become active, as a result of which the grain sizes of ceramics can reach several microns. Currently, with the development of technology, it is possible to obtain polycrystalline materials with grain sizes that are much smaller than the wavelength of visible light, that is, nanostructured ceramics[138].

Optical nanoceramics are synthesized by vacuum sintering of nanopowders (for particles of spherical shape, with surface curvature). On one hand, the surface diffusion of atoms is significantly activated as a driving force for compaction and removal of residual porosity of ceramics at the sintering stage, but on the other hand, the high surface area of nanosize particles contributes to their excessive agglomeration. This leads to the appearance of packing defects in the compact, which prevents uniform compaction of ceramics and lowers the density of the compact. As a result, such defects are preserved in sintered ceramics in the form of light scattering centers, which reduce the optical transparency. A quantitative measure of the agglomeration degree of the nanopowder particles is the magnitude of the fractal dimension, directly related to the structurally sensitive properties of the nanopowders, such as adsorption capacity or sintering activity. For the formation of optical nanoceramics by the method of vacuum sintering, the balance between the sintering activity and the degree of agglomeration of the original nanopowders is optimal.

Optical nanoceramics are characterized by a combination of optical, physical, mechanical and other properties necessary for application in photonics, optoelectronics and scintillation techniques, which are used in the study of outer space.

Currently, technologies of directional energy transfer for remote power supply of spacecraft of long-term operation have drawn high attentions, since the energy supply by their traditional methods no longer provides the required tactical and technical characteristics. In particular, the urgent task is the remote power supply of unmanned aerial vehicles, as well as low-orbit spacecraft, on which the use of classic solar batteries is limited.

In addition, relatively inexpensive micro- and nano-space vehicles are being intensively developed, considered as an alternative to large satellites. A promising direction is the development of low-orbit cluster space systems using remote power supply technologies. The remote power supply of the spacecraft is planned to be carried out with the help of high-power solid-state lasers and high optical quality (minimum beam divergence), ensuring the charging of satellite batteries. At the same time, the laser beam is much more intensive than the solar radiation flux in the case of traditional power supply; therefore, large areas of photodetector components will not be required to receive required amount of energy.

Laser ceramics, as the active medium of solid-state lasers, has a combination of properties that ensures its high competitiveness with respect to both laser glasses and laser crystals[139].

The first laser polycrystalline material (CaF_2) was obtained in 1966 by hot vacuum pressing of powders. Afterwards transparent ceramics with laser properties were synthesized on the basis of Y_2O_3 doped with Nd. ZrO_2, HfO_2 or ThO_2 were used as sintering additives.

Ceramics based on garnet $Y_3Al_5O_{12}$ with a density close to theoretical was obtained in 1984. SiO_2 and MgO were used as additives to improve the sinterability and suppress grain growth. The initial powders were obtained by aerosol spraying of a solution of a mixture of metal sulfates, followed by calcination. The main disadvantage of the $Y_3Al_5O_{12}$ ceramics was a high level of non-selective absorption in the range of 0.25—3.0 cm^{-1}. Due to this parameter, ceramics significantly lost to commercial single crystals and glasses. A breakthrough in the field of laser ceramics occurred in 1995 when ceramics of laser-quality based on garnet $Y_3Al_5O_{12}$ doped with 1.1% of Nd were synthesized. These ceramics were obtained by solid-phase sintering, ensuring a grain size of ~50 μm and a density of 99.8%. After improving the method using vacuum sintering of nanopowders, ceramics were obtained with a transparency equivalent to that of a single crystal.

To obtain a non-porous laser ceramics with a density corresponding to a single crystal with the same composition, it was necessary to ensure as dense packing of particles as possible. High diffusion mobility of atoms in the sintering process and maximum use of internal compressive pressure are the driving force for removing pores during sintering ceramics. The most common method of compaction of raw materials and sintered ceramics is the additional use of external pressure — uniaxial isostatic or magnetic-pulse. In the synthesis of transparent ceramics, the composition and quality of the starting materials are critical factors determining the quality of the products obtained. For the manufacture of transparent ceramics, several approaches are used to the synthesis of the initial precursor powders, considered as follows:

The simple methods are the traditional dry *mechanochemical method*. The essence of the mechanochemical synthesis of precursors is to grind a mixture of oxides in ball mills. To obtain $Y_3Al_5O_{12}$ garnet, high-purity Y_2O_3, Al_2O_3 and Nd_2O_3 oxide powders (>99.99%) with a particle size of 60, 400 and 500 nm, respectively, are mixed in a stoichiometric ratio of garnet and ethyl alcohol, using tetraethyl orthosilicate (TEOS) as sintering aids to ~0.5 wt.%. The mixture is ground by ball milling with alundum balls for 12 h.

The process of grinding a mixture of powders in a ball mill is not a simple crushing of the components. In the collision of grinding balls and powders, the latter particles are destroyed and then "stick together" into various agglomerates.

Mixing particles interact by diffusion and heat exposure upon collision. After a long grinding, the powders consist of individual oxides, but their structure is strongly disturbed and distorted by residual stresses. Moreover, oxide particles are sintered with each other, forming agglomerates with composition close to that of garnet. It is the metastable state of the individual oxides particles after grinding that provides for their more rapid and complete chemical interaction during the subsequent sintering at elevated temperature.

After grinding, the slurry is sprayed and dried to obtain spherical granules of a uniform composition with 50—100 μm in size. The granular powders are pressed in conventional molds and further subjected to Cold Isostatic Pressing (CIP) at a pressure of 98—196 MPa. After double (uniaxial and isostatic) pressing, the density of compacts is 50%—55% of the products.

The green compacts are heated to remove the organic binder and sintered in vacuum (10^{-3} Pa) at a temperature of $T_{sin} = 1\ 750$ °C for 20 h. The obtained ceramics do not contain any optical inhomogeneities, including birefringent regions, for example, visible in garnet single crystals. The heterogeneity of the refractive index in the sample is 10^{-5}, which indicates a high optical quality of ceramics. The technological scheme of the mechanochemical method for obtaining optical ceramics is given in Fig. 8.4[139].

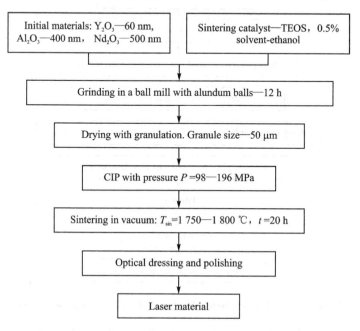

FIGURE 8.4 Mechanochemical synthesis of laser ceramics

Chemical deposition of a mixture of hydroxycarbonates and the subsequent synthesis of nanoparticles are used not only to produce $Y_3Al_5O_{12}$ garnets, but also for transparent ceramics based on other refractory oxides with a cubic lattice—Y_2O_3, $YGdO_3$, Sc_2O_3, Lu_2O_3; Y_2O_3 doped with Er, as well as $Y_3Al_5O_{12}$ doped simultaneously with Cr and Ca[139].

The method of chemical deposition is considered as a promising method for the synthesis of nanopowders, active to sintering, with homogeneous chemical and phase composition both within a single particle and along an ensemble of particles.

For the synthesis of ceramics based on $Y_3Al_5O_{12}$, chlorides or nitrates of Al, Y and Nd, dissolved in water in a stoichiometric ratio, are used as initial solutions. This solution is added dropwise to the deposition—NH_4HCO_3. The deposition is filtered off, washed and dried for two days at 120 ℃. The resulting precursor consists of particles of size 10 nm. Pure garnet powder $Y_3Al_5O_{12}$ with a particle size of 100 nm is obtained by calcining the precursor at a temperature of 1 200 ℃. The technological scheme for the synthesis of yttrium aluminum garnet optical ceramics is given in Fig. 8.5[139].

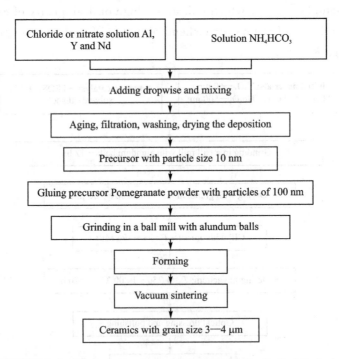

FIGURE 8.5 Hydrocarbonate Synthesis of $Y_3Al_5O_{12}$ Optical Ceramics

Another method of chemical method is the process of decomposition of *citrate gel*, in which the formation of garnet powders occurs at relatively low temperatures. To obtain the initial gel, a mixture of nitrates $Al(NO_3)_3$ and $Y(NO_3)_3$ in a stoichiometric ratio of garnet powders is dissolved in water and a solution of citric acid (2 mol/L) is added at a ratio of 1∶3. The solution is evaporated at a temperature of 80 ℃ and the obtained gel is dried at the same temperature under vacuum for 24 h. Powder for subsequent sintering is obtained by thermal decomposition of the gel at a temperature of 600 ℃ for 6 h. The gel is almost completely decomposed at this stage. In the dynamic heating mode, the complete decomposition of all the intermediate phases, including carbonate, finishes at a temperature of 815 ℃. The beginning of the formation of the garnet phase is at a temperature of 800 ℃ and it becomes the single phase from 1 000 ℃ and becomes highly perfect at 1 200 ℃. With prolonged isothermal annealing, the garnet phase is purified and sufficiently perfect at a temperature of 900 ℃[139].

Nanopowders for laser ceramics are also produced by the method of *self-propagating synthesis*. The essence of the method consists in using self-sustaining redox reactions as a heat source for local heating of the required mixture of starting materials and synthesis of the final product occurring in the reaction zone. The advantages of the method are its simplicity, low-temperature synthesis and the homogeneity of the initial reaction mixture. The reaction mixture is provided by using a gel as an intermediate product, in which a self-sustaining reaction is initiated. Usually, nitrates are used as oxidizing agents, and organic ligands, most often citric acid, are used as reducing agents. When the gel is heated in the first stage (80—200 ℃), alcohol and moisture evaporate. The pyrolysis of organic components and their combustion under the action of an oxidizing agent is initiated at temperatures of 200—540 ℃. In this temperature range, the most significant mass loss occurs (more than 50%) and amount of gaseous products is emitted—CO_2, NO_2, H_2O. The formation of the garnet powders occurs in the temperature range of 600—1 000 ℃ and is accompanied by a weight loss of 5% to 10%. The finish temperature of all chemical processes depends on the nature of the reducing agent and ranges from 925 ℃ for gels with citric acid and up to 960 ℃ for gels with glycine. A feature of this method is to obtain a well-formed garnet phase at an annealing temperature below 1 000 ℃. The synthesized nanoparticles are agglomerates with a diameter of 30—150 nm and a length of 150—500 nm.

Intermediate between the citrate method and self-propagating synthesis is the *Pechini method*. In this method, due to the esterification reaction between citric acid bound to metal ions and ethylene glycol, a polymer is formed in which Al and Y atoms are distributed as evenly as possible. The distribution remains homogeneous after burning off the organic components of the polymer, which results in a decrease in the diffusion length during sintering. The advantage of the method also consists in the low temperature at which nanocrystals with fairly perfect crystal lattice are formed.

During the synthesis of $Y_3Al_5O_{12}$ garnet nanopowders doped with Nd by the Pechini method, a mixture of $Al(NO_3)_3$ and $Y(NO_3)_3$ nitrates is used as starting materials, which is dissolved in an aqueous solution of citric acid and ethylene glycol. The solution is evaporated at a temperature of 90 ℃ until complete polymerization. During calcination, the complete loss of water and decomposition of nitrates occurs in the temperature range of 100—600 ℃. The mass loss of the gel when heated to a temperature of 600 ℃ is ~80%. The formation of the garnet crystal occurs at a temperature below 900 ℃, and the final removal of carbon occurs at a temperature of 1 000—1 100 ℃.

The *method of laser evaporation* of a target by a CO_2-laser beam is used to obtain weakly agglomerated powders. Powders of oxides Y_2O_3 and Nd_2O_3 are used as targets. The resulting powders include two fractions—the main one consisting of spherical particles of average size of 10 nm and a small fraction of spherical particles with a diameter of 0.2—2.0 μm. By sedimentation of the powder in isopropyl alcohol, the fractions are easily separated.

As a result of condensation, oxide nanopowders are obtained in monoclinic modification, which causes the problem of cracking during sintering. To obtain high-quality ceramics, nanopowders are subjected to recrystallization annealing at a temperature of 800—850 ℃, during which the monoclinic lattice is transformed into a cubic one. Annealed powders are pressed by the magnetic-pulse method and sintered at a temperature of up to 1 900 ℃. This technology produces optical ceramics with a residual porosity of 0.03%[139].

Good results in the quality of optical ceramics were obtained using the *sol-gel method* for producing nanopowders of complex composition. This method allows synthesizing nanopowders at low temperatures, which provides nanoscale powders. Al_2O_3, Y_2O_3 and Yb_2O_3 sols are obtained separately and mixed together; during calcination already at a temperature of 850 ℃, a garnet phase is obtained with a narrow size distribution around an average value on the order of 100 nm. The main problem of the sol-gel method is to obtain strongly agglomerated powders. When using this method, the correct choice of precursors is of great importance. Thus, ceramics $Y_3Al_5O_{12}$ of high quality is obtained by using rare earth alkoxides as precursors. When using this method, the garnet phase was the only one at calcination temperatures of 700 ℃. Non-porous ceramics are obtained from the powder that was previously dispersed by ultrasound by pressing and vacuum sintering at a temperature of 1 500 ℃ for 1 h[139].

Currently, work is underway to expand the capabilities of optical ceramics, in particular, to create gradient materials with different refractive index over the cross section, which opens up new possibilities in laser technology and contributes to its further development.

8.5.2 Radio-wave Transparent Materials

Radio-wave transparent materials are structural dielectrics that do not significantly change the amplitude and phase of the radio-frequency electromagnetic wave passing through them ($10^8 - 10^{12}$ Hz), having stable dielectric permittivity and low dielectric loss ($\tan \delta = 10^{-2} - 10^{-5}$, with dielectric constant $\varepsilon < 10$) in a wide range of operating temperatures.

Radio-wave transparent materials are used for the manufacture of fairings and antenna radome protecting the radar equipment from aerodynamic heating and gas (dust) erosion by supersonic or hypersonic gas flow in aviation and rocket-space technology. Depending on the speed of the object in the atmosphere, the maximum level of aerodynamic heating of the surface of radio-wave transparent materials varies from 250 ℃ to 2 500 ℃[20].

The dielectric loss tangent $\tan \delta$ determines the amount of energy absorbed by the material when radiation passes through it. That is, a fairing made of a material characterized by a high $\tan \delta$ value attenuates the electromagnetic signal of the antenna device. In addition, not only the small $\tan \delta$ value is important, but also its stability over the entire range of operating temperatures.

The value of the dielectric constant ε, the tangent of the dielectric loss angle $\tan \delta$ and the wavelength of the transmitting device determine the wall thickness of the fairing. Interference is minimal if the wall thickness is a multiple of half the wavelength of the working radiation. The dielectric constant affects the frequency of the radiation propagated in the radome, and, consequently, the length of its wave. If $\tan \delta$ keeps the minimum value, the wall thickness can be increased by reducing the dielectric constant. The stability of the dielectric constant in the range of operating temperatures of the fairing is also of great importance.

Since during the flight, the radio-wave transparent radomes bear mechanical pressure from the flow and vibration, they must have sufficient mechanical and fatigue strength. In addition, the fairings should be characterized by high heat resistance, which is mainly determined by mechanical strength, elastic properties (E) and the value of the Thermal Coefficient of Linear Expansion (TCLE). As a rule, a decrease in TCLE leads to an increase in heat resistance due to a decrease in thermal stresses.

Glass-ceramic materials are most often used for the manufacture of rocket fairings. They belong to high-temperature materials and are characterized by low TCLE, high operating temperatures, and also possess such a combination of properties, i. e. , homogeneity, low ε and $\tan \delta$, high mechanical strength, wear-resistance and heat resistance, which allow the products based on them to operate in critical conditions. Fairings made of high-temperature glass-ceramic can be successfully used at speeds of $3Ma$ and work under conditions of severe thermal shock (heating rate up to 250—300 ℃/s).

Nowadays, the powder metallurgy method for producing the glass-ceramic materials (sintered sitalls), based on the principles of the technology of making ceramic products, is becoming increasingly common. In this case, the glass of the initial composition is crushed and molded using the methods of ceramic technology, and then heat treated with the purpose of sintering and crystallization.

Powder technology makes it possible to use crystallizing glasses of an extremely wide range of compositions as starting components, including unsuitable compositions for the classical method (from glass melt), i. e. , highly viscous, "short", high crystallization ability. Moreover, in most cases, sintered materials possess a uniform micro-crystalline structure with a high fraction of crystalline phase[140].

The powder metallurgy method can be used to obtain glass-ceramic materials with densely sintered and adjustable open and closed porous structure. Powder technology in comparison with the classical melting technology provides enhanced stability and reproducibility of the properties of these materials. For the powder technology of glass ceramics, the catalyst content can be significantly reduced, sometimes even reduced to zero, since the high specific surface of the particles plays the role of the catalyst, that is, there is a surface crystallization of the particles, which are then sintered into bulk-crystallized material. The glass-ceramics fabrication technology consists of two stages—the preparation of a blank from a slurry of glass powder using a slip method to pour into plaster molds and subsequent sintering-crystallization of the product. The most widely used products are made from the $Li_2O-Al_2O_3-SiO_2$ (LAS) system, in which various additives are used as catalysts to improve the performance properties (TiO_2, Cr_2O_3, Si_3N_4 and others).

The use of new modified technological approaches allows obtaining materials with desired properties. The TSM-983 material based on $Li_2O-Al_2O_3-SiO_2$ with the addition of Cr_2O_3 has increased radio transparency due to the intensification of the sintering process, and the addition of BN reduces the sintering temperature of the TSM-107 material. TiO_2 additives control the value of the dielectric constant of the material TSM-109. Additives Si_3N_4 and SiC help to increase the oxidative stability of radio-wave transparent materials.

Nowadays, spunumen-eucryptite sitalls based on the Li_2O-Al_2O_3-SiO_2 system, characterized by almost zero TCLE value, high heat resistance in the temperature range from -60 ℃ to $+1\,000$ ℃, corrosion resistance and low tan δ values are widely used for the manufacture of fairings. Glass ceramics based on Li_2O-Al_2O_3-SiO_2 (LAS) have proven themselves as the basis of heat-resistant radio-wave transparent materials for the manufacture of aircraft structures. The use of ceramic powder technology allows to vary the phase and chemical composition of glass ceramics and, accordingly, its properties.

An important role in the process of directional crystallization is played by the nature and content of the crystallization catalyst. The introduction of TiO_2 crystallization catalysts into the Li_2O-Al_2O_3-SiO_2 ceramics allows to obtain heat-resistant radio-wave transparent glass-ceramic material, whose main phase is β-spodumene. By varying the additions of TiO_2 and SiO_2, it is possible, over a wide range, to change such characteristics of glass ceramics as the dielectric constant, bending strength, porosity, and, consequently, the service properties in the right direction. These ceramics showed a significant advantage in the combination of physical and technical properties compared to other sitalls and pyrocerams. The technology of shell formation by the method of slip casting into plaster molds (in contrast to the traditional production of the centrifugal casting of glass skewers) provides the best thermal stability characteristics. At the same time, it turns out that this technology is sensitive to the formation of cracks due to moisture-heat shrinkage and gas evolution in the process of molding and firing, leading to uneven physical and technical properties along the height of the shell[19].

In the United States, the most well-known glass-ceramic radio-wave transparent material is Pyroceram 9606 (Corning 9606) from Corning Glass, which is a glass-ceramic system MgO-Al_2O_3-SiO_2 (MAS) with crystallization catalyst TiO_2. The main crystalline phase in such a glass is cordierite MgO-Al_2O_3, which provides low dielectric loss, high resistance to thermal shock, high mechanical strength and resistance to rain erosion. Materials of this class are used for the manufacture of fairings, as well as for sea-based airfield products, due to the high resistance to long-term exposure to marine moisture.

In recent years, most studies of glass ceramics LAS and MAS have been devoted to studying the nucleating action of various crystal catalysts (ZrO_2, TiO_2, and others) at the earliest stages of crystallization, which allows us to create materials with new properties. For example, when using zirconium oxide crystals as catalyst in glass ceramics based on MAS, destabilized solid solutions of high-temperature quartz can be formed. Destabilization contributes to the transformation of the high-temperature form to the low-temperature form during the cooling of the material after crystallization, which leads to the appearance of stresses, and, consequently, to the hardening of the glass ceramics. One of the ideal catalysts for LAS glass ceramics is P_2O_5 since only the lithium disilicate $Li_2O_5Si_2$ phase is formed during crystallization[140].

As the development of aviation, with increasing requirements for speed and maneuverability of aircraft, the requirements for radio-wave transparent materials also increase in particular for dielectric losses ($\tan\delta, \varepsilon$), operating temperature, durability, resistance to water gas and dust erosion during flights at hypersonic speeds. In this consideration, glass ceramics based on $SrO-Al_2O_3-SiO_2$ (SAS) and $BaO-Al_2O_3-SiO_2$ (BAS) are of interest. Such materials contain crystalline phases of monoclinic strontium anorthite and Celsius or their solid solutions, characterized by $T_{mel} > 1700\ ℃$, low $\tan\delta$ and ε values, good strength properties and rather low TCLE value, which make them promising for creating high-temperature radio-wave transparent materials.

Powder technology for producing glass-ceramic $SrO-Al_2O_3-SiO_2$ with the addition of TiO_2 includes glass melting, subsequent grinding and roasting, during which sintering and crystallization take place. In the synthesis of sintered sitalls, the determinant factor is the particle size of the original glass powders. Sintered materials with high density and good mechanical properties can be obtained using powders with a narrow size distribution range with help of a multi-stage sintering mode. This means that when sintering ceramics or non-crystallizing glasses one should strive to reduce the coarseness of grinding to intensify the compaction process. Afterwards, the sintering process at the initial stage prevails over crystallization due to the narrow distribution interval. Using this approach from a $SrO-Al_2O_3-SiO_2$ glass powder with an optimal size distribution of 20—50 μm, radio-wave transparent material was obtained having $p=0.3\%$; $\rho=0.96\rho_{th}$; $R_{m,ben}=130$ MPa; $H_\mu=9600$ MPa. At a temperature of 1 100 ℃ glass ceramics based on $BaO-Al_2O_3-SiO_2$ has $\varepsilon=6.55—7.0$ and $\tan\delta=(8—25)\times 10^{-4}$, based on $SrO-Al_2O_3-SiO_2$ has $\varepsilon=6.16—6.77$ and $\tan\delta=(11—50)\times 10^{-4}$[140].

Since alkali-free ceramics is refractory and the cooking temperature of the original glass reaches 1 600—2 000 ℃, in most cases the preparation of glass-ceramic materials based on them is carried out using ceramic technology. In addition, the synthesis of the initial powders is carried out using the sol-gel method. This method allows obtaining nanopowders with a uniform distribution, which after the sintering stage are transformed into a monolithic glass-ceramic material, which is characterized by higher mechanical properties as compared to the glass ceramics obtained using traditional ceramic technology. However, there is a problem of aggregation of the initial powders, which can lead to residual porosity of the material. The problem is solved by increasing the pressure at the compaction stage.

Radio-wave transparent materials with $T_{mel} < 2\ 000$ ℃ are powdered ceramics based on refractory compounds, such as oxides, nitrides, borides, possibly reinforced with fibers of refractory oxides. Modern technologies allow us to obtain a large range of properties due to the manufacture of materials with various porosities (from dense to highly porous) and by modifying them with various additives to obsess special properties (with high radiating ability, adjustable dielectric constant and high thermal erosion resistance). In addition, gradient materials are of particular interest, in which a combination of layers with different densities and chemical compositions is possible, which makes it possible to tune the properties of materials.

A series of materials based on Si_3N_4 was developed by Ceradyne, Inc. The company produces several types of antenna radomes from materials such as fused quartz, IRBAS, Ceralloy 147-31N and Ceralloy 147-01EXP, which are listed in Table 8.7. Since materials used in rocket systems are subjected to high mechanical and thermal loads, their dielectric properties are required to be stable up to a temperature of 1 400 ℃. Fused silica SiO_2 has a low thermal conductivity and TCLE, and shows high thermal shock resistance. However, its strength properties are low, and the operating temperature does not exceed 1 100 ℃. The strength properties of Si_3N_4 are much higher than those of the fused SiO_2 and the dielectric constant is stable with increasing temperatures. However, the value of tan δ rapidly increases above the temperature of 1 000 ℃. Ceralloy 147-31N (Si_3N_4) has a higher purity than IRBAS, better strength properties and more stable electrical properties. Ceralloy 147-01EXP is reaction-sintered Si_3N_4. Its strength characteristics are higher than those of SiO_2, but lower than those of Ceralloy 147-31N. The advantage of this material is the stability of its dielectric properties up to a temperature of 1 400 ℃.

TABLE 8.7 Properties of radio-wave transparent materials of ceradyne, Inc.

Properties	Fused quartz	IRBAS	Ceralloy 147-31N	Ceralloy 147-01EXP
Composition	SiO_2	Si_3N_4	Si_3N_4	Reaction-sintered Si_3N_4
$\rho/(kg \cdot m^{-3})$	2 000	3 180	3 210	1 800—2 500
$R_{m,ben}/MPa$	43	550	800	180
E_{ben}/GPa	37	280	310	50—200
$\lambda/[W \cdot (m \cdot ℃)^{-1}]$	0.8	20	25	6
ε	3.3	7.6	8.0	4—6
$\tan \delta$	0.003	0.002	0.002	0.002—0.005

The company Secretary of the Air Force (USA) has developed a product from Si_3N_4 with regions of different density. Monolithic fairing consists of the anterior rears and transitional parts. The front part has a density of $\rho = 750$—$1\,000$ kg/m³, the back part is $\rho = 1\,600$—$2\,000$ kg/m³, and the transitional part is characterized by a variable density. The method of producing a fairing includes filling and compacting the mold with Si, Si_3N_4 particles and fugitive pore-forming additives, removing the green compact from the mold, sublimating the fugitive additives to organize the porous structure, and interacting with nitrogen to convert the porous structure into a structure containing "whiskers" based on α-Si_3N_4.

The company Secretary of the Navy (USA) manufactured an electromagnetic window made of ceramics with composition Si_3N_4, where $AlPO_4$, ZrP_2O_7 compounds or their mixture were used as binder.

Of all materials currently used, sintered Boron Nitride (BN) has the best dielectric properties: $\tan \delta = 0.001$ at a temperature of 1 500 ℃, and a change of $\varepsilon \leqslant 3\%$ with increasing temperature to 1 350 ℃. However, BN has low strength and hardness, so it is used as a component in composite materials.

A durable, erosion-resistant dielectric ceramic material obtained by powder metallurgy technology has been developed. The composite contains a discrete phase BN, distributed in the Al_2O_3-matrix. Boron nitride particles have a size of 3—5 μm and are able to withstand thermal loads without destroying the material.

The material of the AlN-BN system are used for radio-wave transparent windows of high-speed aircraft and the method of its production are patented by General Electric company (USA). The material contains (0.001—35) wt.% BN, $\leqslant 2$ wt.% Al_2O_3, B_2O_3 or SiO_2, the rest is AlN. The method of obtaining structural ceramic products includes mixing the starting materials in the form of a powder mixture, its formation and compaction. The dielectric properties of the composite with the content (5%—35%) BN slightly change in the temperature range of 0—1 400 ℃.

Loral Vought Systems company (USA) proposed a method for obtaining material from an antenna radome that includes mixing a powder mixture containing (20%—60%)Si_3N_4, (12%—40%)BN and (1%—20%)oxygen-containing sintering agent (MgO, CeO, Al_2O_3, Y_2O_3 or their mixture); molding the mixture and hot pressing. The produced radio-wave transparent material has ρ = 2 400—2 900 kg/m^3, ε = 4.5—7.0, and tan δ = 0.001. The material is stable at the temperature of 2 000 ℃ and shows a tendency to ablative ablation only at a temperature of 2 500 ℃.

Currently, considerable experience has been gained infabricating structures and parts from quartz ceramics. The main disadvantage of ceramic and fiber-reinforced composite materials based on quartz ceramics is their rather low strength. Therefore, such materials were initially used in aircraft construction for the manufacture of radio-wave transparent windows of telemetry antennas and small-sized fairings of onboard communication systems. For this type of structures, strength $R_{m,ben}$ = 20—40 MPa is sufficient, and the installation of such structures on an aircraft does not cause difficulties and does not require special dimensional accuracy.

Radio-wave transparent materials based on Si_3N_4, BN, AlN compounds are being developed for operation in the temperature range above 1 000 ℃. Of particular interest are the works on the synthesis of ceramics, including those made from a gradient composition, from a material under the common name "Sialon" (SiAlON), consisting of various compounds in the silicon-aluminum-oxygen-nitrogen system.

Si_3N_4 is a unique material with a combination of excellent radio-wave transparent, thermal, chemical and mechanical properties. However, it, like other nitride compounds, tends to oxidize at high temperatures with the formation of SiO_2. In addition, under the influence of high temperatures, diffusion of impurities to the surface occurs, which reduces the strength characteristics of the material. It is possible to produce such a gradient material in the "SiAlON" system, in which the performance characteristics would approach to silicon nitride and the oxidation resistance close to silicon oxide. The diffusion of impurity atoms in this case is designed to be blocked by selected intermediate layers. Given so many possible options in this system, the task is not simple and requires studying.

8.5.3 Magnetic Materials

Magnetic materials are materials whose relative magnetic permeability is substantially higher than 1. Magnetic materials are divided into ferrimagnetic and ferromagnetic. There are hard magnetic materials, soft magnetic materials and magnetic materials for special-purpose, i. e., magnetodielectric, magnetostrictive materials, superconducting magnets, magnetics with a rectangular hysteresis loop, thermomagnetic materials, and ferrites for super-high-frequency engineering.

Hard magnetic materials are magnetic materials with low magnetic permeability, large coercive force ($>$10 kA/m) and large hysteresis losses. These magnetic materials are of the greatest interest in aircraft design from the point of view of powder metallurgy.

The emergence of new materials for permanent magnets based on Rare-Earth Metals (REM) and 3d-transition metals has made a revolutionary leap in the field of hard magnetic materials. In the 70s of the 20th century, K. Strnat patented the technology for producing pressed powder magnets from the $SmCo_5$ compound. The stoichiometric alloy was comminuted to an average particle size of $\sim$5 μm (almost single crystal), whose coercive force is 5—10 times greater than previously achieved values. Afterwards the powders were compacted in the presence of a magnetic field to obtain a crystalline texture. In the production of permanent magnets, there was a transition from casting technology to powder metallurgy. At the same time, not only the technology for achieving high magnetic characteristics changed, but also the ideas about the nature of the highly coercive state and the processes of magnetization reversal have changed fundamentally, and a new opportunity for producing of superpower magnets and devices based thereon has appeared[113].

The manufacturing techniques of such magnets are typical for powder materials with one point—the powders are compressed in the presence of a magnetic field (or the mixture is kept in a magnetic field to orient the particles before pressing). Hard magnetic materials based on REM-Co possess the greatest crystallographic magnetic anisotropy of all known materials, which ensure the possibility of obtaining magnets with record values of the coercive force in terms of magnetization and specific magnetic energy therefrom[141].

The REM-Co alloys are divided into two groups—AB_5 and A_5B_{17} (where A is REM, B is a transition metal). The AB_5 group is characterized by the highest crystallographic anisotropy, with low saturation induction values and the Curie point compared to the A_5B_{17} group. The mechanical and physical properties of the

REM-Co are given in Table 8.8.

Magnetic materials based on intermetallic compounds $SmCo_5$, $PrCo_5$ and $Sm_xPr_{1-x}Co_5$ have the superior properties (for $x < 1$). Intermetallic melting is carried out in metallic crucibles or by crucibleless melting in arc or induction furnaces in a protective atmosphere of argon or vacuum. To improve the homogeneity of the structure, cooling of the melt in the crystallization range is carried out slowly. Then, the obtained ingots are ground to powders with a particle size of 5—20 μm. When grinding, the surface hardening occurs, accompanied by an increase in the concentration of defects and micro-stress, which leads to an increase in coercive force[59]. The magnetic characteristics of the powders increase when their surface is treated with dilute acids (they remove a non-uniform surface layer) followed by non-electric deposition of Ni, Co, Sn or other metals. The powders or mixture of powders are treated in an external magnetic field of 2.0—2.4 mA/m in strength to obtain a magnetic texture and pressed (the magnetic field is sometimes applied at the moment of pressing directly) in molds at a pressure of ~500 MPa. Then, pressing of the blanks (porosity of 30%—40% after uniaxial pressing) is performed in a hydrostat at a pressure of 2 GPa (single or double pressing). The blanks pressed in this way have a porosity of 2%—3%. Compacts are sintered in an inert atmosphere at a temperature of $T_{sin} = 1\,000$—$1\,100$ ℃ for 1 h, and after sintering, the density of magnets is almost theoretical[59].

TABLE 8.8 Mechanical and physical properties of REM-Co compounds[141]

Compound	B_r/T	$(BH)_{max}/(kJ \cdot m^{-3})$	$\rho/(kg \cdot m^{-3})$
YCo_5	1.060	216.0	7 600
$LaCo_5$	0.909	165.2	8 030
$CeCo_5$	0.770	118.4	8 550
$PrCo_5$	1.200	288.0	8 340
$SmCo_5$	0.965	186.4	8 600
$(MM^*)Co_5$	0.890	158.4	8 350
Y_2Co_{17}	1.250	312.0	—
Ce_2Co_{17}	1.150	264.0	8 730
Pr_2Co_{17}	1.380	380.0	8 560
Sm_2Co_{17}	1.200	288.0	8 720
Tm_2Co_{17}	1.300	256.0	9 240

* MM—Mish Metal, mixture of REM. Values are given for technical MM composition 54.5 Ce-26La-13Nd-5Pr(cerall).

To obtain magnets from ternary alloys $Sm_{0.5}MM_{0.5}Co_5$, $Sm_{0.5}Ce_{0.5}Co_5$,

$Sm_{0.5}Pr_{0.5}Co_5$ and others, liquid-phase sintering is used. The stoichiometric alloy is mixed in a jet mill to a particle size of 6—8 μm. Alloy powders enriched with Sm (40Co-60Sm) are added to the powders to a content of 63 wt. %Sm. The mixture is placed in elastic tubes and the magnetic field with a strength of 125—750 mA/m and the particles are oriented. Afterwards, the powders are evacuated and compressed on a hydrostat at the pressure of 1.38 GPa. Compressed blanks are sintered in high-purity argon at the temperature of 1 100 ℃, which exceeds the melting point of particles from a samarium-enriched alloy[59].

In addition to high magnetic characteristics, powder magnets have 10 times higher strength than cast ones ($R_{m,com}$=620 MPa), but at the same time, they are 4—10 times more expensive. The magnetic properties of the sintered magnets based on REM-Co are given in Table 8.9[113].

TABLE 8.9 Magnetic characteristics of alloys based on REM

Alloys	Composition		B_r/ T	H_c/ (kA·m^{-1})	$(BH)_{max}$/ (kJ·m^{-3})
	Sm	Sm+Pr			
KS37	36.0—38.5	—	0.77	540	110
KS37A	35.0—38.6	—	0.82	460	130
KSP37	—	36.0—38.5	0.85	520	130
KSP37A	—	36.0—38.5	0.90	500	145

As modern research has shown, the magnetic characteristics of permanent magnets depend not only on the anisotropy organized in the production process, but also on the grain size of the intermetallic microstructure. The technology of production of permanent magnets based on physical phenomena—Hydrogenation – Disproportionation – Desorption – Recombination (HDDR), allows to obtain magnets with a microstructure of 50—100 nm, by regulating the performance of these processes. HDDR are phase transformations that occur in intermetallic compounds when interacting with hydrogen during heat treatment in hydrogen and vacuum at temperatures up to 950 ℃[142].

For the production of permanent magnets using hydrogen treatment, the sintering operation is carried out in a regulated manner—first in hydrogen, then in vacuum. This approach, in addition to grinding the grains of intermetallic compounds, makes it possible to reduce the sintering temperature of the magnets from 1 000—1 100 ℃ to 950 ℃; it also contributes to the preservation of nanoscale grains[143]. The proposed technology was tested on $SmCo_5$, Sm_2Co_{17}, $Nd_2Fe_{14}B$ magnets.

Technological scheme of the production of permanent magnets is given in Fig. 8.6.

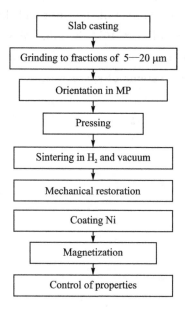

FIGURE 8.6 Technological scheme of permanent magnets

A schematic diagram of the organization of the anisotropic properties of magnets from nano-intermetallic compounds using the example of $SmCo_5$ is shown in Fig. 8.7[143].

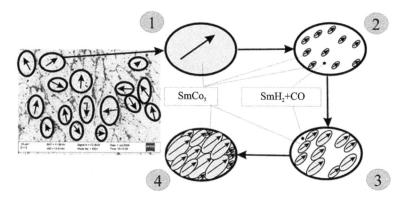

1—The production of an oriented single crystal in the process of grinding and orientation in a magnetic field;
2—Hydrogenation with the provision of a residual magnetic phase;
3—Dehydrogenation and growth of the magnetic phase;
4—The formation and growth of new grains in compliance with the initial orientation

FIGURE 8.7 Stages of pabricating magnets with anisotropic properties by HDDR

In aerospace technology, such magnets are used for alternators with a power of more than 100 kV·A. For operation in space vehicles, a magnetic bearing was designed, excluding the direct contact of the rotor and the stator at a rotor speed of 24×10^3 r/m, at a torque of 5×10^{-2} N·m. Magnets produced by powder metallurgy technology are used in magnetic recording systems in the heads of rockets and other electronic systems of spacecraft.

Chapter 9 New Technologies in Powder Metallurgy

9.1 Nanomaterials and Nanotechnologies

At the present stage of development, the equipment, in particular, the equipment in aerospace, has reached such point that classical materials and approaches to the technology of manufacturing of parts have come to saturation and further progress in terms of improving performance characteristics leads to significant costs with minimal progress. It is raised an awareness of the need to involve new non-traditional materials and technological approaches in the production of technology. In this regard, nanotechnologies offer almost unlimited opportunities for the development of aviation and astronautics, the development of fundamentally new constructive approaches to the technologies and the resolution of the problems of human-machine interaction.

The principal advantages of nanomaterials and nanotechnologies are the creation of completely different operational qualities of materials as compared to existing ones. This will allow creating more efficient equipment, primarily in terms of weight efficiency, with lower energy consumption and environmental impact.

The special properties of materials at the nanoscale level are demonstrated by the fact that particle sizes (molecules, clusters, grains, domains) become commensurate with the scales of such physical quantities as the average electron path in metals, the size of domains in ferromagnets, the phonon path in crystals, and effective exciton size in semiconductors. The properties of nanomaterials are basically dominated by quantum effects.

Nanomaterials are very diverse. Take nanopowders as an example. They can be used as an independent material, for example, as catalysts for chemical processes. Solid high-strength materials can be obtained by pressing and sintering of nanopowders. As fillers, nanopowders can be part of composite materials that have unique mechanical, tribological, thermal, or electrical characteristics and their combination.

A promising direction in the use of nanomaterials is the blending of nanopowders to conventional powder mixtures, which can significantly improve the properties of the products. When adding 0.1—0.5 wt.% Ni nanopowders to the mixture of Ni and Fe powders, the porosity of the products decreases by 4%—7% while simultaneously reducing the sintering temperature by 150—200 ℃. For nickel-molybdenum steels, the replacement of carbonyl nickel by nickel oxalate nanopowders increases the strength of the product by 1.5 times, and their plastic properties by 4 times. The addition of nanopowders in the composition of 0.5Ni-(0.5—1.0)Cu-0.3C to the powders of 17Cr-2Ni steel allows to obtain powder metallurgy steels with impact strength KCU = 1.1—1.5 MJ/m^2, which is approximately equal to cast steel and 1.5 times higher than that of the deformed steel. With the introduction of such an additive, the porosity of steel decreases by 5%—6%, and the hardness increases 1.5 times, reaching 1.2—1.6 GPa[113].

The development of nanotechnology is largely associated with the discovery, study, and commencement of practical use of three carbon nanostructures—fullerenes, carbon nanotubes, and graphene (see Fig. 9.1).

 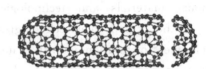 

(a) Fullerene molecules C_{60} (b) The carbon nanotubes with end section   (c) Graphene

FIGURE 9.1 Structures

Fullerenes, which are also informally called buckyballs, are spherical molecules consisting of 20 or more carbon atoms, forming a net shell of five- and hexagonal-rings. Each molecule contains $2(10+n)$ carbon atoms (n is the number of hexagons). The most stable spherical shapes are formed if the pentagons are separated from one another by no more than one hexagon (the rule of isolated pentagons). Fullerenes got their name in honor of the architect Buckminster Fuller, who invented similar structures for use in architecture.

The most studied material is fullerene C_{60}, which was discovered in 1985 in experiments on laser evaporation of a graphite target. The surface of the C_{60} molecule is a polyhedron consisting of 20 hexagonal and 12 pentagonal faces (see Fig. 9.1(a)). The diameter of the C_{60} molecule is ~0.7 nm, and the diameter of its internal cavity is 0.5 nm.

In addition to the nanoscal, fullerenes have a unique set of properties: solubility (in benzene, 2.5 mg/mL), properties of a semiconductor and a superconductor, the ability to withstand large voltages or selectively block light. The fullerene derivatives form compounds with atoms, molecules, ions, radicals, and so on. Moreover, the latter can be added both outside the fullerene molecule (exofullerenes) and inside (endofullerenes)[144].

The structure of the fullerene molecule is interesting because a cavity is formed inside such a carbon "ball" into which atoms and molecules of other substances can be introduced, which provides, for example, the possibility of safe transportation of the latter.

Fullerenes are already being used in the production of rechargeable batteries, the possibilities of creating optical shutters on their basis, as well as devices for recording information and solar cells. C_{60} molecules are also suitable for use as additives in the creation of various nanomaterials, rocket fuel additives, and antifriction coatings.

A *carbon nanotube* is a monoatomic graphite layer, 1—5 nm in diameter. Fig. 9.1(b) shows a schematic representation of the end parts of the nanotube, which are closed with hemispherical caps that look like half a fullerene. For the first time, carbon nanotubes were discovered in 1991 with the high-temperature destruction of graphite electrodes with an electric arc ignited between them. Then the formation of carbon nanotubes was observed with laser evaporation of graphite.

Nanotubes, like fullerenes, can be filled with atoms and molecules of another substance. An example of carbon nanotubes filled with magnetic nanoparticles is given in Fig. 9.2[145].

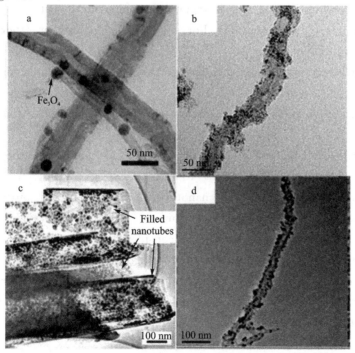

FIGURE 9.2 Carbon nanotubes with filled magentic nanoparticles

In addition to single-walled carbon nanotubes, there are multi-walled carbon nanotubes, which are several single-walled nanotubes nested inside each other. The diameter of multi-walled nanotubes reaches 20—25 nm, and the distance between the layers is 0.34 nm, which corresponds to the distance between carbon atoms layers in graphite. The length of nanotubes produced by the electric arc method and laser evaporation usually does not exceed 10—100 μm. Significantly longer nanotubes are produced by chemical vapor deposition from hydrocarbons.

Nanotubes have very good mechanical properties. For single-walled carbon nanotubes $R_m = 50$—150 GPa, which is tens of times higher than the strength of steels. Since the density of nanotubes is quite low ($\rho = 1\,300$—$1\,400$ kg/m^3), the specific strength of the material made of nanotubes is fantastic. Young's modulus of nanotubes is two times higher than that of conventional carbon fibers. Carbon nanotubes have a fairly high plasticity ($A = 10\%$—15%). Under the action of mechanical stresses above the critical value, the nanotubes do not "tear" and do not "break", but simply rearrange.

Carbon nanotubes also have unique electrical and thermal properties associated with the characteristics of their structure. Depending on the particular folding scheme of the graphite plane (chirality), nanotubes can be both conductors and semiconductors. Chirality is a structural characteristic of nanotubes, determined by the relative orientation of the hexagonal graphite mesh with respect to the longitudinal axis of the nanotube. For single-walled nanotubes, there are three possible coagulation options:

① Armchair—two sides of each hexagon are oriented perpendicular to the tube axis;

② Zigzag—two sides of each hexagon are oriented parallel to the axis of the tube;

③ Chiral—each pair of sides of the hexagons is oriented to the axis of the nanotube at an angle different from 0° to 90°.

The electronic properties of nanotubes can be changed purposefully by introducing other substances inside the atoms.

Fullerenes and carbon nanotubes are considered as potential nano-additives of polymer composites. In carbon fiber reinforced polymers, it becomes possible to carry out a nanostructural modification of the interface between polymer matrix and fiber reinforcements (a critical region in composites during fracture) by dressing the fiber with a fullerene solution. By adding carbon nanotubes to the binder, a modification of the supramolecular structure of the matrix is achieved, which makes it possible to eliminate the "friability" and reduce the proportion of free volume. Thus, it is feasible to increase the packing ratio of the matrix and the composite as a whole. With the help of functional derivatives of fullerenes, a bulk reinforcing cage can be organized in the polymer matrix, which would alleviate the negative effect of anisotropy, a well-known deficiency of layered composites.

As part of aviation materials science, it is advisable to use fullerenes and nanotubes when developing the following types of polymer-based aviation materials, such as adhesives, sealants with high adhesive and cohesive strength, rubber with improved wear resistance properties, paint coatings with increased adhesion and resistance to erosion. The addition of fullerenes or carbon nanotubes significantly increases the wear resistant properties of metal-fluoroplastic antifriction materials, and reduces the coefficient of friction.

One of the methods for producing nanoscale diamonds, fullerenes, nanotubes and coatings based on them, for use in aerospace engineering, is the method of detonation synthesis. Detonation synthesis of carbon nanostructures is based on the unique features of carbon atoms which join together in giant molecules and form spatial structures with strong bonds. The high density of mobile carbon atoms, which are formed at the peak of the over-compressed detonation wave, is a necessary condition for the formation of carbon nanostructures in detonation synthesis. Fast-flowing (10 ns) processes of energy and matter conversion in a detonation wave contribute to the efficient formation of allotropic carbon structures from atoms, first in the form of energetic precursor, and after their rapid cooling and quenching, forming nanodiamonds, fullerenes and carbon nanotubes.

Graphene is an isolated monoatomic layer of graphite, that is, a layer of atoms located in the nodes of a hexagonal two-dimensional crystal lattice (see Fig. 9.1(c)). The properties of graphene are unique in many ways. It has very high thermal conductivity at room temperature ($\lambda = 5 \cdot 10^3$ W/(m·K)), which is ~15 times higher than that of copper and 1.5 times higher than the same indicator of nanotubes. At present, graphene is considered the most promising substitute for silicon as a semiconductor material for nanoelectronics.

Graphene is extremely resistant to microscopic high-speed bullets. A single-carbon layer is 8 to 10 times better at dissipating the impact energy than steels with similar thickness. Therefore, graphene is ideal for protecting satellites, interplanetary ships and probes from meteorites and "space debris".

The main areas of application of nanotechnology in aviation include the followings.

1. Structional materials and materials with special purposes

This area includes:

① ultralight, ultrastrong, corrosion-resistant, wear-resistant, fretting, erosion-resistant and heat-resistant;

② heat-resistant, being able to reduce the weight and dimensions of structures at the same time;

③ adaptive, including with shape memory and properties of superelasticity;

④ hydrophobic and self-cleaning to combat the icing problem.

2. Membranes and coatings

This area includes:

① defect-free surface layers of nanostructured materials applied in the zone of mechanical stresses concentration in order to increase strength, resistance to crack nucleation, allowing to increase the durability and endurance of structures;

② antibacterial coatings and structional materials (internal cavities of fuel tanks, pipelines, various surfaces, as well as components of aircraft subject to biodeterioration);

③ micro- and nano-porous membranes for thermomolecular pumps;

④ coatings to reduce visibility in the radar range and the creation of infrared camouflage systems.

3. Adhesives and lubricants

This area includes: electrically conductive adhesives, paints, new types of lubricants (including solid ones) for engines and other high-loaded friction units of an aircraft (landing gear, wing control mechanisms, etc.), which increase tribological characteristics of the units and the temperature limit of their use.

The strength issues of aircraft structures can be solved by creating aluminum and titanium alloys reinforced with carbon nanotubes, creating new composite materials using nano-components. The use of nanocoatings in the areas with maximum mechanical stress will increase the durability and reliability of aircraft structures.

The use of modern materials is limited there, that an increase in strength leads to a decrease in ductility. The advantages of materials with a nanoscale structure is an increase in plasticity and a decrease in residual stresses. For example, titanium-based nanoalloys provide an increase in the relative elongation in the range of 20%—35% with an increase in strength of ~ 2 times. Due to the use of nanotechnology, an increase in the strength of the primary structural materials is predicted by 1.5—3 times.

Currently, in solving problems of strength and ductility control of structures, the optimal properties of units and devices can be achieved by using composite materials. The scientific and technical background for the use of nano-modified composite materials in aircraft structures is related to the characteristics of carbon nanomaterials, primarily nanotubes. It is impossible to use carbon nanotubes directly in structures but using them as a reinforcement in composite materials is very effective not only because of its high strength, but also due to an increase in the interface area compared to reinforcements with normal sizes. When the size of the reinforcements varies from micro- to nano-meter, the interfacial area with the same volume fraction can increase by more than 30 times, thereby reducing the stresses from the existing external loads. The main obstacle in the practical use of nanomaterials, which is to be overcome, is their thermodynamic instability and tendency to grow and coarsen. Obtaining bulk nanocrystalline materials with a uniform structure (without pores, microcracks and other defects) will ensure the stability of properties, more accurately determine the bearing capacity of the material and, as a result, increase the safety factor and reduce weight of the structure.

The durability of the materials with stress concentration in a hole is determined by the growth process of existing microcracks located on the edges and the surface of the hole. When assembling the airframe of a large passenger aircraft, several million holes are drilled, which act as stress concentration regions. When drilling holes, technological microdefects (microcracks) are inevitably formed. Under cyclic loading of the structure during operation, these defects develop, significantly reducing the lifetime of the aircraft.

It is assumed that the application of defect-free layers of nanostructured material on the edges or surface of the hole will lead to an increase in strength and durability due to the complex effects, namely:
- when introducing nanolayers it is possible to self-heal existing microcracks on the surfaces and edges of the hole—the stress concentration regions;
- the presence of the applied layer can shift the zone of the substrate material with microcracks to a region with lower stress concentration.

This layer can protect the underlying material from environmental influences and thus prevent damage by corrosion cracking or corrosion fatigue.

The most important task of aerodynamics is to reduce the resistance of the aircraft. Reducing the aerodynamics resistance of the aircraft, performing a long cruise flight, leads to a significant reduction in fuel consumption. The value of the aerodynamics resistance of the aircraft can be divided into components having a different physical nature: pressure resistance and friction resistance.

The friction resistance can be reduced by 80% when the flow regime in the boundary layer on the entire surface of the aircraft changes from turbulent to laminar. In this regard, it is necessary to create devices for laminarization of the boundary layer. With their help, laminar flow can be obtained in 30% of the wing area and aircraft empennage, which will lead to a decrease in friction resistance by 25% and full resistance by 10%.

Nowadays, there has been developed a system that can be used to control the boundary layer by means of a flexible tape with ∼10 mm wide and 0.1—1.0 mm thick, glued to the surface of an airplane wing parallel to its leading edge. The tape must be connected to a computer that controls the operation system in accordance with the characteristics of the boundary layer at the installation site.

The device reduces the magnitude of the unstable disturbances passing over the tape by 1—2 times. As a result, the boundary layer must remain laminar at a distance of 20—50 cm behind the tape (provided that the laminar flow is in front of it). A large number of sensors and actuators (actuators that introduce small disturbances to the boundary layer) should be placed on the tape surface. As actuators, micron cavities can be used with a piezoelectric crystal, whose vibrations form a vortex pair equal in amplitude and opposite in the direction of rotation to the perturbations of the boundary layer. A predetermined connection between the sensor signals and the action of the actuators is carried out by a microcircuit located inside the tape. The typical sizes of actuators and sensors should be about 100 μm. For their manufacture it is necessary to use nanotechnology. Such devices can significantly improve the efficiency of aircraft.

It is also possible to use nanotechnology to directly change the properties of aerodynamic surfaces, both physical and mechanical, in particular, the surface roughness, which largely determines the friction resistance.

A very topical issue is the protection of aircraft against icing, which is directly related to flight safety and the complexities of the anti-icing system itself and its maintenance. The presence of even a small amount of ice on the leading edge of the wing can lead to a significant decrease in the lift coefficient, which negatively affects the conditions of take-off and landing. Many examples of accidents due to the icing of the wing and the loss of controllability of the aircraft were known in history.

Nanotechnology to create hydrophobic (and superhydrophobic) surfaces can significantly advance in solving the problem of icing[146-147]. To do this, it is necessary to create such a hydrophobic surface (possibly a film or coating) that would prevent the formation of an ice crust on the edge of the wing and tail of the airplane, and if it appeared, the ice can be completely eliminated by heating or deforming the shape[148]. For example, Cao et al. studied the anti-icing capability of the nanoparticle (silica)-ploymer composite coatings on the Al plate, as shown in Fig. 9.3. In the test of anti-icing properties in naturally occurring "freezing rain", little ice was shown on the side with the superhydrophobic composite (with 50 nm particle) while the untreated side is completely covered by ice. Similarly, for a commercial satellite dish antenna (see Fig. 9.3 (e) and (f)), no ice was covered on the one-half coated with the superhydrophobic composite whereas the other half untreated was completely covered by ice. They found that the anti-icing capability of these superhydrophobic surfaces is dependent strongly on the size of particles.

(a) SEM image of a particle-polymer composite made of 20 nm silica particles

(b) Untreated side of an aluminum plate after the natural occurrence of "freezing rain"

(c) Treated side of the aluminum plate coated with a superhydrophobic composite after the "freezing rain"

(d) SEM image of a particle-polymer composite made of 50 nm silica particles

(e) Satellite dish antenna after the freezing rain. The left side is untreated and is completely covered by ice, while the right side is coated with the super-hydrophobic composite and has no ice

(f) Close-up view of the area labeled by a red square in (c), showing the boundary between the coated (no ice) and uncoated (ice) area on the satellite dish antenna[130]

FIGURE 9.3 Transmission electron microscopy image and SEM images of anti-icing superhydrophobic coatings[147]

To reduce the visibility of aircraft in a wide wavelength range, it is necessary to create nanostructured stealthy materials. Such coatings will provide a significant increase in the effectiveness of military equipment. When designing aircraft with a reduced level of visibility, the result does not always depend only on the layout scheme and on giving the aircraft a low-reflective shape. The many limitations and requirements imposed on aircraft in terms of aerodynamics, strength, controllability, and so on, do not allow the full use of many layout-constructive ways to reduce visibility. In some cases, there is a need to manufacture components of aircraft from structural absorbing materials or absorbing coatings on these materials[149].

To reduce the visibility of the body of aircraft, different types of coatings are used. The use of paint coatings is not effective since after a few months of their application, the surface becomes almost matte. In this regard, in the development of new coatings, one of the main requirements is to ensure stable reflective properties. This is possible in case of creating coatings resistant to erosion wear.

The requirements for absorbing materials include:

① light enough not to increase the weight of the device;

② having a small reflection coefficient in a wide range of wavelengths of electromagnetic radiation;

③ stable and reliable in operation under various climatic conditions;

④ easy to maintain.

When reducing the surface of the absorbing components to critical values, the magnetic and electrical properties of the surface receive qualitative changes. Thus, when using the latest advances in nanotechnology, it becomes possible to create absorbing structures based on new physical principles. For this purpose, for example, layers of polymer matrix with fillers in the form of one-dimensional (nanotubes, fibers, microwires), two-dimensional (flakes, films) and three-dimensional (fullerenes, microspheres) additives can be used. Layers of such composite materials demonstrate qualitatively new radio-absorbing properties[145].

Another area of application of nanotechnology is related to the development of instrumentation (sensors for measuring pressure, friction, temperature) based on nano- and MEMS-technologies.

MEMS-technology (technology of micro-electromechanical systems) is one of the most revolutionary technologies of the 21st (twenty-first) century, allowing to integrate computing processes into the physical structure of matter itself[150]. The main goal of research in the MEMS-technology is to combine computing with physics. Micro-electromechanical systems are integrated devices made on a semiconductor substrate (most often silicon) and containing mechanical structures, sensors, actuators and electronic components. Typical dimensions of micromechanical units (system components) are in the range from 1 μm to 100 μm, while the size of a MEMS crystal chip reaches values from 20 μm to 1 mm.

At present, the state of the aircraft is monitored by general integral characteristics and during the flight it is impossible to obtain accurate information about the state of the structure and the flow conditions at a specific point of the aircraft. It is possible to realize global monitoring of aircraft flow parameters by the application of nano- and MEMS-sensors, which can provide accurate information about the state of the structure and flow, as well as measurements of the factors that influence the state in terms of improving the working conditions of the aircraft bearing units[151]. Some examples of MEMS devices for aerospace applications are shown in Fig. 9.4[152].

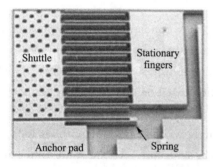
(a) SEM photo of a MEMS shear stress sensor, showing the inter-digited fingers, shuttle, anchor pad and spring

(b) Array of high temperature MEMS heat flux sensors

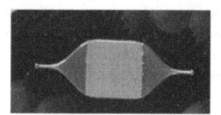
(c) MEMS-based microcatalytic reactor

FIGURE 9.4 Examples of MEMS devices

In studies of dynamical similarity, the main difficulty lies in creating a model that corresponds to actual objects and accurately reproducing all the necessary elastic-mass properties. In this case, one often has to deal with overloading the model with respect to a full-scale object. The overload ratio can range from 1.3 to 2.0. This means that the mass of the model can be exceeded by 30%—100%. Inaccuracy in modeling the properties of a full-scale object will adversely affect the reliability of the experimental results. This can be avoided by using materials with a higher elastic modulus, for example, carbon nanocomposites with a set of properties that are superior to those of full-scale materials.

Like aviation, astronautics is one of the most promising branches of technology, where nanomaterials and nanotechnologies can be used on a large scale. In the coming future, nanomaterials and nanotechnologies are expected to be increasingly used to create new types of space technology, which will make it possible to change approaches to the design of spacecraft and will play a huge role in the implementation of large-scale space projects.

With the help of nanotechnology, composite materials of two types are created: nanocomposites and nano-nanocomposites. The first type is composites in which nano-sized reinforcements are used, and the matrix is not nanostructured, and the second type is those having a nanostructured matrix strengthened with nano-inclusions.

With the development of nanotechnology, new opportunities have appeared for improving the properties of ceramics and composites used for rocket and space technology units that are exposed to high temperatures (above 1 200 ℃). The addition of 0.25% carbon nanotubes to ceramics based on silicon carbide leads to an increase in strength by 30% or more.

The reinforcement of composites with single-walled carbon nanotubes helps to obtain a material that is both durable and wear-resistant. At the same time, it is possible to obtain a composite with high strength and high crack resistance—a combination not typical for ceramic composites.

The use of nanocomposites in rocket and space technology will allow to:

① improve specifications by increasing operating temperatures;

② dramatically increase the work resource;

③ reduce weight due to high specific strength and simplified design;

④ increase the reliability of the rocket engine due to the avoidance of forced cooling.

At present, composite nanomaterials with the following characteristics are being developed for rocket and space technology: $R_m^{1\,000} > 500$ MPa; $H_\mu \sim 30$ GPa; $E > 280$ GPa, durability $> 20\,000$ h at $T = 1\,500$ ℃. The production of a turbine from nanocomposites for the transport energy module is planned by the method of slip casting, and for the gas duct is planned by the method of slip casting of films, followed by hydro- or gas-static sintering, Spark Plasma Sintering (SPS) or electric pulse sintering.

The National Aeronautics and Space Administration (NASA), Air Force Research Laboratory (AFRL), Rice University in the United States and a number of other research centers are successfully developing ceramics-based nanomaterials for production of turbines, gas ducts, combustion chambers, and so on, using SiC,

BN, Si_3N_4, Al_2O_3, ZrO_2, HfC as initial components. The NATO Research and Technology Organisation is developing nanocomposites for microturbines. In Japan, multilayer Al_2O_3/ZrO_2 composites strengthened with SiC nanoparticles were obtained.

Honeywell, University of Houston and University of California are developing Si_3N_4-based nanocomposites with different fraction of nanoparticles as reinforcements, which are obtained by SPS or Hot Isostatic Pressing (HIP) sintering at low pressure. To obtain composites based on ZrO_2 with low thermal conductivity ($\lambda = 0.37-0.4$ W/(m·K) in the range of $T = 550-1\,050$ °C) fullerene C_{60} additives are used.

In India, a layered nanocomposite based on ZrO_2 with a strength of up to 630 MPa and a hardness of HV=13 GPa has been created, whose size is governed by the number of layers. In Brazil, an Al_2O_3/ZrO_2-based nanocomposite with microhardness $H_\mu = 22$ GPa was obtained by SPS-sintering.

A promising direction in nanotechnology is the creation of materials and systems that simulate biological objects and use the principles of their operation (*biomimetic and bioinspired materials*). Such materials and systems can be built only on the basis of inorganic materials or with the use of biological objects, and during their creation, self-assembly processes can also be implemented. With the use of nanomaterials, unique components are being fabricated that are actively introduced into electronics, computer technology, measuring systems, and so on. Such components, of course, will find their application in the onboard equipment of aerospace.

Due to their high electrical and thermal conductivity, nanotubes can be used as connecting wires in chips with tight packing. The significant specific surface area of carbon nanotubes makes it possible to consider them as material for creating ultra-miniature high-capacity compensators (supercapacitors), which can be used not only to create electronic devices, but also as storage components in the power supply system of spacecraft and other devices in which powerful pulses of energy are required.

When analyzing the ways of developing the components for building onboard systems of future spacecraft, it is necessary to recognize the importance of two promising areas—spintronics and photonics. Spintronics, in contrast to traditional electronics, operates with electron spin and the electron's associated magnetic moments. Photonics uses quanta of light as information, which fundamentally provides a number of advantages over traditional charge electronics in terms of speed, noise immunity, power consumption and heat dissipation. Nanotechnologies and nanomaterials created with their help open up completely new possibilities for the development of various sensors—sensors for recording and determining parameters for a wide range of physical objects. At the same time, the majority of nanosensors differ from existing sensors not only by their small size, but also by significantly better service characteristics.

Nanotechnologies also allow the creation of highly sensitive sensors of mechanical effects, accelerations, and electromagnetic fields. A feature of such sensors is the presence of moving parts, which allows to receive an electrical signal in response to a mechanical effect, and vice versa—a mechanical response to an electric effect.

The combination of electronic and electromechanical nanodevices in a single chip, including devices for controlling the flow of liquid or gas, allows to create complex functional systems, called Lab-On-a-Chip (LOC). The result of the development of the considered design principle is the creation of a "spacecraft on a chip".

In the first spacecraft, thermoelectric materials were used, which converted the heat generated by radioactive decay into electricity as batteries for space probes. A significant drawback of such materials is their low efficiency. Currently, a highly efficient material has been developed that can convert radioactive radiation itself into electrical energy. The new nanomaterial is multilayer carbon nanotubes filled with gold and surrounded by lithium hydride. Radioactive particles hitting gold, knock out high-energy electrons from it, which pass through nanotubes and enter lithium hydride, from which they continue their way along electrodes. Such material makes it possible to obtain a 20 times higher return from radioactive decay compared with thermoelectric materials. New thermoelectrics can become compact and convenient sources of energy for spacecraft.

Analysis of the state of research in the field of nanomaterials allows us to make a forecast about the practical application of several classes of nanomaterials. First of all, new structural materials based on nanoparticles and nanotubes will dramatically reduce the weight of aircraft without reducing functionality. Another successful direction is nanoelectronics using nanotubes and other nanostructures. In future, it is planned to create molecular computers and biocomputers that are resistant to the effects of space radiation, develop biosensors using the effect of "molecular recognition", create sensor networks for diagnosing and monitoring the environment in the vicinity of the Earth and other planets. Another direction is the creation of nanorobots system capable of performing large-scale work, including those aimed at improving the environment and human exploration of outer space.

Nanotechnologies are considered as one of the key points in the implementation of promising space projects. A program has been developed for the application of nanotechnology in the development of the Moon, Mars, Venus and the creation of large satellite systems for scientific research and monitoring of the Sun and the space environment.

Until 2035, continuous improvement of the main spacecraft systems is expected through the use of nanotechnologies and nanomaterials. This will allow in the coming years to use the "constellations" of small satellites to monitor the Earth, the near-Earth space environment and the Sun, and later to commission similar "constellations" for studying deep space and build space telescopes with a large aperture. The robots created with the use of nanotechnologies and nanomaterials will play an important role in the study of Mars, Venus and the giant planets of the Solar System, in the vicinity and on the surface of which the devices must operate under extreme conditions. The implementation of such large space projects as the construction of habitable bases on the Moon and manned mission to Mars is also not possible without nanotechnologies and nanomaterials, ranging from the use of structural and functional materials to the use of various devices and systems providing diagnostics of the health status of cosmonauts.

The successful implementation of programs for the development of space nanotechnologies and the introduction of nanomaterials into space technology is associated with the solution of a number of tasks to create new unique nanomaterials with the properties necessary for their application in space technology. On the other hand, it is necessary to study the behavior of nanomaterials and products based on them in the conditions of outer space.

To achieve these goals, it is necessary to:

① create physical and mathematical models that adequately describe the structural features of nanomaterials and the effect mechanisms of outer space factors on nanostructures;

② develop technological and test facilities of a new generation for the production of nanomaterials and the study of properties evolution under the influence of space environment factors;

③ prepare and conduct space experiments on testing nanomaterials[146].

The main difficulty that must be overcome when creating nanomaterials is the thermodynamic stability of matter in solid state. Materials tend to a minimum of free energy, which is determined, in general terms, by the ratio of volume to surface area, and therefore to an increase in volume with an optimal geometric shape.

9.2 Additive Manufacturing

Additive Fabrication or Additive Manufacturing (AF/AM) are terms meaning the method of manufacturing products with the "adding", in contrast to the traditional mechanical manufacturing "subtracting excess" material from the bulk of the workpiece. These terms are used in parallel with the phrase *Rapid Prototyping* (or RP-technology) but have more general meaning and are the development of RP-technology[153].

The basis of the "rapid prototyping" method (stereolithography) is based on the combination of two modern tools: computer and laser. A necessary condition for the manufacture of parts by this method is the presence of its computer-based three-dimensional model (solid form) with the help of which the laser is controlled. The second side of the question is how and what a ray will affect. The origin of the idea is lithography and there is one of the first names "stereolithography". Having a three-dimensional model, any part can be divided into many sections with the necessary pitch, and the laser, reading a two-dimensional drawing, can quickly draw these sections, starting from the base to the top of the product. If these sections (flat plates) are glued together, we can get the exact model of the prototyping part. The first version of this technology (Laminated Object Modeling (LOM) process) involved the use of adhesive paper, each layer of which along the contour was cut off by a laser. As a result, a paper-based exact model was obtained that could be immediately used to model the product and produce molds with a small number.

The next stage in the development of RP-technology is the formation of a model from polymeric materials. The basis of the method is the polymerization of liquid organic monomers under the influence of a laser beam (see Fig. 9.5). In the initial state, the platform (table) of the manipulator (3) is immersed in the monomer liquid to a depth of Δt. The layer of liquid above the platform will be molded in one pass of the laser, which scans the surface of the liquid, according to a given program, turning it into plastic. After the scan is completed, the platform is lowered again to a depth of Δt and the scan continues according to the program of the

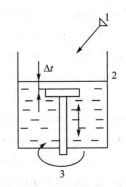

1—Laser; 2—Monomer bath; 3—Manipulator

FIGURE 9.5 Scheme of laser polymerization

next layer and so on until the solid plastic model of the part is obtained. In this way, it is possible to obtain a model of a part, whose dimensions are limited by the size of the platform and the depth of the bath. The accuracy of the dimensions is determined by the accuracy of the laser beam scanning. Such products could be used as models for the subsequent manufacture of metal parts by the method of shape casting or as plastic structural parts with complex shape. The next stage in the development of RP-technology was the selective laser sintering of metallic powders.

Rapid prototyping in the modern sense is part of the AF/AM technology, which is prototyping using layered synthesis methods. Additive technologies cover all areas of product synthesis, be it a prototype or a serial product. Thus, in the course of its development, additive production has evolved from a tool for rapid prototyping to a set of efficient technologies for the manufacture of parts. Additive technologies are related to the technologies of the 21st century. Additive methods of creating parts is the transition to the sixth technological mode (*waves of innovation*, *techno-economic paradigm*).

The traditional production chain of parts manufacturing in general form is a sequence of operations—ingot manufacturing, hot deformation to obtain a semifinished product, preliminary mechanical processing to obtain a billet, heat treatment, final mechanical processing for the production of a part. In powder metallurgy, metal is first produced in the form of a powder, which is then pressed into a special form in a billet, which is sintered at high temperature in furnaces, after which the final part is mechanically processed. In the case of additive manufacturing, instead of traditional multi-stage process chains, a shorter one is used—the powder is immediately sintered into the final part of a given shape and exact dimensions. At the same time, the designer has the opportunity to conduct a comprehensive testing of the assembly or part, including checking its strength properties, kinematics of functioning, durability, and adjusting the design.

The essence of AF/AM technology consists in the layered synthesis of products by fixing layers of model material and their successive connection in various ways depending on the tool (method) used: sintering, gluing, polymerization. The ideology of AF/AM technology is based on digital technologies, which are based on a digital description of the product—its computer model (CAD model). When using AF/AM technology, all stages of project implementation from idea to materialization (in the intermediate or final version of the finished product) are in a "friendly" technological environment—a single technological chain, where each operation is performed in a digital CAD/CAM/CAE system (*CAD*: *Computer-*

Aided Design; CAM: *Computer-Aided Manufacturing*, materialization on technological equipment by means of numerical control; CAE: *Computer-Aided Engineering*, calculation and analysis by means of special software). In practice, this means a real transition to "paperless" technologies, which means traditional paper and technological documentation are not required in principle when for the manufacture of parts.

Fig. 9. 6 gives examples of the industry share of the additive manufacturing market in 2016 and 2017. The contribution from the aerospace industry increased from 2016 to 2017 and arrived 18. 2%. The use of additive technologies provides:

① production of complex-shaped and unique parts (for example, "detail in detail" or parts gradient in composition and properties, etc.) without using machining process and expensive specialized equipment;

② elimination of the human factor in the manufacture of parts (the construction of the part is made fully automatically);

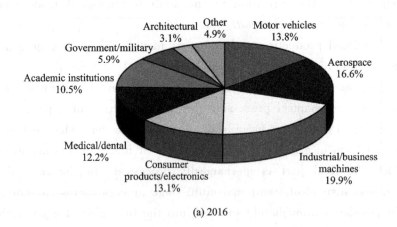

(a) 2016

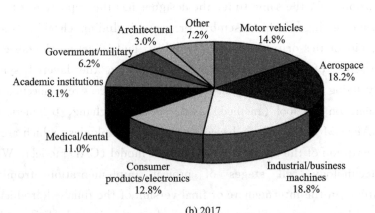

(b) 2017

FIGURE 9.6 Additive manufacturing market share by industry in 2016[154] and 2017[155]

③ reducing the mass of the part by reducing the thickness of the walls, components, creating cellular and other structures (the so-called bionic design);

④ ability to replace block parts on reinforced structures by adding additional stiffeners;

⑤ ability to create integrated parts for one technological cycle;

⑥ possibility of applying the necessary material directly to the desired areas without the use of fasteners-screws, nuts, flanges;

⑦ expansion of functionality due to improved design, and not only due to the microstructure of the material;

⑧ management of the physical, mechanical and chemical properties of the product being created;

⑨ production of materials with higher metallurgical uniformity (products have higher mechanical properties compared to casting ones and are close to those obtained by hot deformation);

⑩ absence of casting defects and stresses in the part;

⑪ increase of the profitability of the production of small series and exclusive options.

The advantages of additive technologies should also include the possibility of instant transfer of digital models to anywhere in the world, which allows, if necessary, to immediately organize their local production on a global scale, which is especially interesting for the repair of aircraft[156-157].

In addition to the obvious advantages in the speed and quality of manufacturing products, additive technologies have an important advantage in terms of environmental protection and, in particular, the emission of greenhouse gases and thermal pollution. AF/AM technology has great potential in reducing the energy costs of creating a wide variety of products[158].

In the aircraft industry there is a phraseology "Buy-to-Fly" ratio, which is explained as "the attitude of what I bought to what flew". According to various data, this ratio, depending on the complexity of the part and the material of the semi-finished product, is 15:1 or even 20:1 (sometimes 50:1). The use of additive technologies reduces this ratio to $(1.5-2):1$, which depends on the materials[159]. AF/AM technology allows solving the following tasks in the aerospace industry—reducing the mass of products, which leads to a reduction in fuel consumption and air emissions, reduction of design time and testing of aerospace equipment, reduction of the production cycle (including assembly of units), as well as innovative use of high-strength metal alloys[153].

There are two variants of additive technologies that differ in the method of layer formation. As shown in Fig. 9.7, in the first case (powder bed technology, see Fig. 9.7(a)), pour the powder onto the work platform firstly and level it with a leveling device in a uniform layer with the required thickness. Afterwards powder particles are fused by the laser or electron beam in accordance with the current cross-section of the CAD-model. The second variant of additive technologies (powder feed technology, see Fig. 9.7(b)) consists in the deposition of feedstock directly to the point where energy is supplied and where the fragment is currently being built[160].

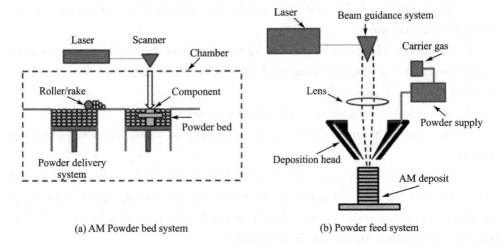

FIGURE 9.7 Schematics of AM powder bed system and powder feed system[160]

For additive technologies, fine powders of various alloys are used as feedstock. The general requirement for powders for AF/AM technology is the spherical shape of particles, due to the maximum packing density of spherical particles in a given volume of the part being manufactured. Spherical powders also have maximum fluidity, which provides minimal resistance in the system of supplying material to the place of the layer formed[159].

Powders for additive technologies should have a stable chemical composition and do not contain impurity gases (oxygen, nitrogen, hydrogen) that cause the formation of porosity during the formation of a layer. One of the parameters characterizing the powder for use in additive technologies is the average particle size and the fraction of such particles in the powder, donated by D_{50}. For example, $D_{50} = 30$ μm, which means that 50% of the powder particles do not exceed 30 μm.

The use of fine powders for additive technologies has a number of advantages. The smaller the value of D_{50}, the smaller the construction step can be set, the more precisely the small components of the part can be made, and the smoother the surface can be obtained from the constructed part. On the other hand, in the process of building, a large amount of energy is instantly injected into the laser/electron spot zone, the melting process proceeds very rapidly, the metal boils and some of the metal flies out of the construction zone. If the powder has a very small particle size, then in the process of building light fine particles, the fine particles will "fly out" from the melt zone, which will lead to the opposite result—increased roughness of the part and high microporosity. In order to avoid the particles "flying out" from the melt zone depositing on adjacent, already formed sections, a directional flow of air or protective gas is introduced inside the working chamber. If the optimum speed is exceeded, the building material may also be carried away from the building zone. Therefore, when working with fine powders ($D_{50} <$ 10 μm), low-power and, consequently, low-performance lasers are used. Such powders are mainly used for the manufacture of microdetails, which cannot be made in any other way. The main methods for producing powders for AF/AM technology are gas (including plasma), vacuum and centrifugal atomization[159-161].

During gas atomization, the raw material is melted in the melting chamber (usually in vacuum or inert atmosphere) and then discharged through a special sprayer device, in which the flow of liquid metal is interrupted by a stream of inert gas under pressure. To obtain fine powders ($D_{50} = 10$—40 μm), VIM (*Vacuum Induction Melting*) atomizers are most commonly used in AF/AM technology. In VIM atomizers, the melting chamber is evacuated to minimize the contact of the molten metal with oxygen and nitrogen. The technology of producing powders using machines for vacuum melting is called VIGA (*Vacuum Induction melt inert Gas Atomization*), that is, "the technology of gas spraying of metal melted in a vacuum chamber by induction heating"[162].

The EIGA (*Electrode Induction melting inert Gas Atomization*: induction melting of an electrode with spraying gas) technology is a type of gas atomization and is designed specifically for spraying chemically active metals and alloys based on them (Ti, Zr, Hf, V, Ir, Nb, Mo and other). Melting of these metals in ceramic crucibles is practically not feasible even under vacuum conditions due to the interaction of the melt with the crucible material. In the EIGA technology, the rods (*feed stock*-raw material) after preliminary melting, according to the adopted technology, pass inductive remelting in the form of electrodes and are sprayed without using a crucible. Melting is performed by lowering a slowly rotating electrode into a ring inductor. Drops of metal are dumped from the electrode into the nozzle system and sprayed with an inert gas. Most often, this method produces powders of refractory and active metals and alloys—TiAl, FeGd, FeTb, Zr, Cr and superalloys based on Fe, Ni and Co[159,163].

Another type of gas atomization is the *plasma atomization* technology, in which the electrode is melted using plasma. Using plasma technology, spherodization of powders is carried out, which consists in melting particles of irregular shape in an inert gas plasma jet, as a result of which they acquire a regular spherical shape[164].

When vacuum atomization in the melting chamber located below the atomizer (soluble gas atomization), an overpressure is generated by gases (H_2, N_2, He), which dissolve in the melt. During atomization, the metal, under the action of pressure in the melting chamber, comes up to the nozzle apparatus, leaving the spray chamber, which is pre-evacuated. The resulting pressure drop induces an intensive force of the dissolved gas to the surface of the melt droplets and explodes the droplets from the inside, ensuring the spherical shape of the particles and the fine powder characteristics.

Technologies of centrifugal atomization are very diverse. High practical interest has been paid to those that allow to obtain powders of reactive and refractory metals. *REP (Rotating Electrode Process)* technology involves spraying a melt created by an electric arc between a bar of metal being sprayed and a tungsten electrode. The main advantage of this method is the complete elimination of contact with the crucible and filling device.

A variant of this technology is the *PREP (Plasma Rotating Electrode Process)* technology, in which the rod is melted using a high-speed flow of ionized inert gas. The REP and PREP technologies, as well as the plasma atomization technology, are used to obtain highly pure spherical fine powders[165]. In recent years, intensive work has been carried out on the use of electron-beam atomization methods for obtaining very pure spherical powders[166].

The development of AF/AM technology in the aerospace industry was promoted by a significant expansion of the range of metal-powder materials. If at the beginning of the century no more than 5—6 types of powder compositions were present on the market, now dozens of brands of powder materials are offered ranging from structural and special steels to high-temperature alloys and rare metals.

Additive technologies that are already used or have practical application prospects include Selective Laser Melting (SLM), Selective Laser Sintering (SLS), Selective Heat Sintering (SHS), Electron Beam Melting (EBM), Laser Metal Deposition (LMD) and Spray Forming(SF) technologies.

1. Selective Laser Melting (SLM) technology

Selective laser melting is an additive manufacturing technology that uses CAD data as a source of information and energy in the form of high-power laser radiation to create 3D parts from metal powders by melting small powders[167].

The technology of layer-by-layer laser alloying of powders was developed in Europe and the USA in the 90s of the 20th century. The Air Force Research Laboratory's Materials and Manufacturing Directorate (Materials & Manufacturing Directorate AFRL/ML) in the USA conducted a large-scale research project to introduce these technologies to create new materials used in the aerospace industry. Initially, the goal was to develop a method of layer-by-layer laser alloying of powders *Laser Additive Manufacturing* (*LAM*) proposed by AeroMet. Rapid Product Innovations (RPI) and EOS (Electro-Optical Systems),GmbH gave their name to this process—*Direct Metal Laser Sintering* (*DMLS*). The name SLM is used by British experts.

The essence of the SLM technology is as follows. Firstly, a thin layer of high-quality metal powder is evenly distributed on the platform. Next, the powder is melted under the selective influence of lasers controlled by two high-frequency scanners, forming a 2D slice of the model. Subsequently, the platform is lowered vertically by the amount of the melted layer and again the melting of the next 2D slice of the model is repeated.

The process of manufacturing parts using SLM technology is carried out in a working chamber filled with an inert gas. This avoids the oxidation of the powder, which allows the use of materials such as aluminum and titanium. Melting is performed using a laser beam directed along the X and Y axes by two mirrors with a high deflection speed. The power of the pulse laser beam is sufficient to melt the powder into a homogeneous material.

Selective laser melting is used to manufacture products of complex geometric shapes and structures with thin walls, hidden cavities and channels. Also, by using the SLM technology, a uniform connection of two parts can be obtained without cracks, pores and holes. With SLM technology, it is possible to simultaneously produce several parts, whose dimensions are limited only by the dimensions of the working chamber of the installation.

2. Selective Laser Sintering (SLS) technology

Selective laser sintering is an additive manufacturing technology in which a mixture of materials with different melting points is subjected to heat treatment (laser). It can be used to produce materials with complex structure. The flexibility of the technology is achieved through direct computer control of the process, and unlike traditional methods of manufacturing parts that require machining, three-dimensional parts are made directly by layer-by-layer sintering of powders[168-169].

SLS technology has similarities with SLM technology[170]. The difference lies in the amount of energy applied to the layer of the metallic powders to be consolidated. In case of SLS technology, the energy input is less (solid-phase or liquid-phase sintering), and in case of SLM technology, the energy is sufficient to completely melt and build a monolithic part. These differences determine a more accurate geometry of the part in the first case and higher mechanical properties in the second case.

The power of lasers in SLS installations is no more than 50—100 W in the Infrared Radiation (IR) range (wavelength is usually 10.6 μm, but there are installations with a wavelength of 1.06 μm). Inert gases, i.e., itrogen or argon is supplied to the chamber in order to avoid oxidation of the powder during heating.

Currently, there is a large number of materials produced by SLS technology:

① single-component Fe, Ti, Cu;

② multicomponent and intermetallic systems (Ni, Al, Ti, W, their carbides and intermetallic compounds, alloys—INCONEL alloy 625, Ti-6Al-4V, Mo-Cu, Cu-Ni, Cu-Sn, Cu-Sn-Ni, Fe-Cu, WC-Co, TiC-Ni/Co/Mo, TiCN-Ni, TiB_2-Ni, ZrB_2-Cu, Fe_3C-Fe);

③ low and high alloy steels;

④ structural (ZrO_2, Al_2O_3, Al_2O_3-$Al_4B_2O_9$, Al_2O_3-SiO_2-B_2O_3, Si_3N_4-Al_2O_3, Ti-Zr-Al/Al_2O_3, Ti-ZrO_2, ZrO_2-Al_2O_3-Al, ZrO_2-Al-Y_2O_3) and ferroelectric (Pb(Zr_xTi_{1-x})O_3, $Bi_4Ti_3O_{12}$, $Li_{0.5}Fe_{2.5-x}Cr_xO_{19}$) ceramics.

Products of the required form obtained by SLS technology can be subjected to additional processing, in particular, annealing and infiltration (impregnation with low-melting metals or polymers) to improve their performance. Hot Isostatic Pressing (HIP) is utilized to eliminate residual porosity for the products obtained by additive technologies, improving the performance properties of products[171]. SLS technology has become widespread throughout the world and continues to thrive thanks to its ability to easily and quickly obtain the most complex objects, including those with a gradient composition, using data embedded in a computer.

3. Selective Heat Sintering (SHS) technology

Selective heat sintering is an additive manufacturing technology that uses directional heat from special lamps instead of a laser, which ensures the simultaneous formation of a layer (unlike most AF/AM technology, by which parts are being manufactured gradually by beam scanning). A special lamp is covered with a mask, and thus it becomes possible to selectively interact with the source material[172].

One of the main advantages of SHS technology is the possibility of selecting an IR radiation of two materials for a certain wavelength, one of which will transmit heat and the other will reflect. It is also possible to reverse the option when it is possible to choose such a wavelength of IR radiation, at which a particular material will absorb or reflect radiation.

4. Electron Beam Melting (EBM) technology

Electron beam melting is one of the layer-by-layer manufacturing technologies, by which not a laser beam but an electron beam is used as the energy source for melting metal[173]. Advantages of EBM technology include:

① providing a high concentration of energy in a narrow focal spot;

② prevention of the formation of oxides and nitrides, and ensuring high mechanical characteristics of products due to the melting of the metal in high vacuum.

Compared with laser technology, EBM technology:

① has more efficient energy, lower installation and maintenance costs;

② provides higher building speed of making the product;

③ provides the structure which is denser, as close as possible to the structure of an ordinary metal due to the complete melting of the powder in the melt spot.

In the aerospace industry, Arcam (Sweden) machines are used to implement EBM technology. The dimensions of the produced products are 200 mm × 200 mm × 350 mm (or diameter is 300 mm, height is 200 mm) and the beam power is up to 4 kW. The thickness of the construction layer, depending on the material, is 0.025—0.25 mm. EBM technology makes parts from powders of corrosion-resistant steels[174], Inconel alloy, titanium and cobalt-chrome alloys[175].

5. Laser Metal Deposition (LMD) technology

Laser metal deposition is an additive manufacturing technology that uses high-power laser radiation to melt metal powders delivered to the weld zone by a laser beam through a special nozzle (see Fig. 9.8(a))[176]. Laser radiation travels along the central part of the nozzle and is focused by one or more lenses. The coordinate table (or nozzle) moves in the $X-Y-Z$ directions to fuse each layer of the 3D object. Powder material is supplied and distributed around the nozzle circumference independently or by means of a transport gas. Gas is also used to shield the surfacing zone from oxygen to the air, reduce oxidation and ensure high wettability of the surface.

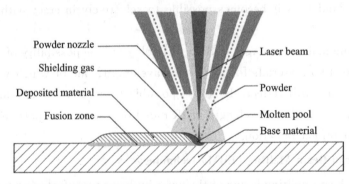

(a) Schematic diagram of the laser metal deposition setup

(b) Five-axis LMD printer manufacturing an IN718 helicopter engine combustion chamber

FIGURE 9.8 Additive manufacturing

In the production of parts using LMD technology, the composition of the product can be changed dynamically and continuously, depending on the goals of the manufacturing. For example, the Direct Metal Deposition (DMD) installation of the company ROM uses CO_2-laser, whose deposition rate ranges from 14 cm^3/h to 147 cm^3/h with a layer thickness of 0.3—0.6 mm. The DMD 505 installation has a 5-axis manipulator and the "rotation" option, the size of the working space is 2 000 mm × 1 060 mm × 610 mm, using 4—5 bunkers for various powders (the ability to create functionally gradient layers and products) with a 5 kW laser. The company TWI has succeeded in producing an IN718 helicopter engine combustion chamber, featuring an overhanging flange using 5-axis system (see Fig. 9.8(b))[177].

The technology of *Laser Powder Surfacing* (*LPS*) is demanded primarily for the restoration and hardening of the surface, but can also be used for manufacturing bulk products, including those with a gradient composition, and hence properties[178]. Such products, in which the surface has special functional properties (wear and corrosion resistance or heat resistance) can be manufactured in one technological cycle.

6. Spray Forming (SF) technology

Spray forming is a new technology, located at the junction of various traditional technologies[179]. It combines the processes of casting (melting metal) and spraying metal (powder metallurgy). The essence of the technology lies in the layer-by-layer sputtering of the metal on the substrate and the "growth" of the workpiece for subsequent hot deformation and mechanical processing. The metal is melted in the melting chamber, then poured from a special sprayer, where the metal jet is atomized into droplets by a stream of inert gas. Metal particles with size of 10—100 μm are deposited on the substrate, forming the body of the workpiece (pseudo ingot). In the end, the preform is produced in layers, which "unites" spray forming technology with AF/AM technology (see Fig. 9.9)[180]. The platform with the substrate can perform a reciprocating motion (then a billet in the form of a strip is obtained) or a rotational motion relative to the flow axis of the sprayed metal (to obtain a cylindrical billet). Rotating movement around an axis perpendicular to the flow axis allows to obtain blanks of the "ring" or "pipe" type. Obtaining different shapes of blanks allows to reduce costs at further thermal processing.

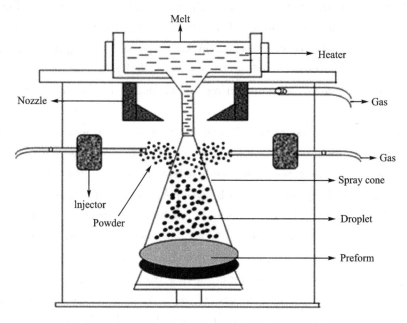

FIGURE 9.9　Spray forming technology scheme

Spray forming technology was developed primarily for the manufacture of critical parts of aviation and space from special alloys with increased requirements for structure, chemical composition and properties. Unlike castings, the billets obtained by the spray forming method have a high homogeneity of the micro- and macro-structures of the metal. The uniform distribution of ingredients across the billet array and the uniformity of the structure are the main advantages of this technology. The disadvantages include the relatively high loss of material during atomization and deposition of the metal on the substrate, as well as the relative complexity of process control.

A feature of the spray forming technology is the presence of micropores in the metal structure, whose appearance is associated with the "capture" of gas molecules during atomization and precipitation of metal particles and with the "loose" sintering of particles during crystallization[181]. Porosity is eliminated by HIP, followed by deformation treatment (forging or rolling) and selection of the sprayed gas, which is absorbed into the metal during the subsequent hot treatment as an interstitial solution (for example nitrogen in iron-based alloys). Spray forming technology opens up wide possibilities for creating gradient materials due to the possibility of applying materials of different chemical composition through two or more spray nozzles.

With the help of spray forming technology, it is possible to create new materials with unique properties, as well as a variety of coatings. Composite materials have already been obtained in which the matrix base is reinforced with ceramics up to 15% by volume, which makes it possible to significantly increase rigidity and wear resistance. Such materials are made by injecting ceramic particles into an atomized metal stream during metal deposition using spray forming technology.

A unique material was obtained in the Al-Si system, in which it is possible to increase the Si content to 70%. Such alloys cannot be fundamentally obtained by casting because of the catastrophic embrittlement of the alloy during its slow crystallization. By changing the ratio between Al and Si, it is possible to produce alloys with a given TCLE (thermal conductivity over a wide range of temperatures). These alloys have prospects for use in microwave devices, film radiators used in aerospace engineering, as well as for automotive engine pistons.

At present, AF/AM technology is moving from the category of experimental and individual works to the category of industrial sites. A number of firms producing aviation, rocket and space products have already been equipped with many devices for the manufacture and repair of parts using AF/AM technology. For example, Pratt & Whitney established an AF/AM technology center. At Boeing's AF/AM technology center, more than 22 thousand parts of 300 items for 10 types of aircraft are manufactured and the nomenclature is constantly expanding.

According to General Electric (GE), in the coming years, about half of the units of power turbines and aircraft engines will be manufactured using AF/AM technology. The fuel injector manufactured by this company using AF/AM technology combined 20 components that were previously manufactured separately, which reduced the weight of the product by 25% with cost reduced by 33% and service life extended in 5 times. GE Aviation has simplified the design and AM technology has realized a reduction from 855 parts produced using conventional manufacturing into a dozen parts. The weight was reduced and the fuel efficiency was improved up to 20%, and 10% more power was achieved thanks to the simplified design with AM technology. A bearing support and sump were redesigned to consolidate 80 parts into one. Also, a 20-part nozzle was consolidated into a single AM unit and the weight was reduced by 25%[182]. Similarly, Airbus simplified a 126-part hydraulic housing tank to a single AM part as shown in Fig. 9.10[183]. The Russian companies VIAM and Aviadvigatel are working to manufacture parts from nickel alloys with the help of AF/AM technology in a closed cycle—starting from the production of charge material.

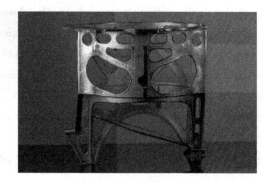

(a) AM hydraulic reservoir rack from Airbus consolidating 126 parts

(b) Consolidated design into one part

FIGURE 9.10   Consolidation by AM

NASA has been successfully using 3D printers with laser sintering technology for powder materials to create critical parts for spacecraft engines. If earlier rocket builders took several months to create one component, 3D printing of the same component takes several days. Automation of production using 3D printers allows people to print parts for spaceships of virtually any complexity and any size.

Using additive DMLS technology the Space Exploration Technologies Corp. (SpaceX) manufactured the SuperDrago engine chamber for the reusable Dragon spacecraft. It is the feature of the DMLS technology that allows the rocket engine to produce up to 7 257.48 kg of thrust, which will ensure the maneuverability and precision of control in orbit and during its passage through the Earth's atmosphere.

DMLS technology is widely used in the segment of satellite construction. Airbus Defense and Space has successfully applied this technology to optimize the design of the brackets connecting the satellite body with solar panels and radio antennas. The parts created on the EOSINT M280 installation of EOS complied with the requirements of the technical conditions—resistance to large external voltages (up to 20 kN) at a temperature difference of −180 ℃ to +150 ℃. In addition to achieving high performance characteristics, the use of 3D printing has allowed a 20% reduction in production costs and labor-intensive production of the brackets.

The European Space Agency announced the launch of the AMAZE project, which aims to print metal parts for spacecraft, aircraft engines and rockets on a 3D printer. The most ambitious goal of the project is to create a space satellite assembled entirely from printed components.

Additive production is a critical component of the success of bases on the moon or mars. An extremely promising technology is required to make Ti, Al or Si powders from minerals mined on the Moon, and then use them to create the necessary infrastructure. Digital technologies allow astronauts to use materials that they find wherever they go and build everything they need there regardless of tools, equipment and infrastructure. Since it is absolutely impossible to ensure the supply of a sufficient number of spare parts for repair or replacement, additive manufacturing will become a fundamental condition for future space travel.

9.3 Severe Plastic Deformation of Sintered Billets

Severe Plastic Deformation (SPD) is a new generation of metal forming processes, ensuring the formation of specified microstructures in them.

The qualitative change in the properties of metals with severe plastic deformation is associated with the formation of structural components with nanometer size, including grains with diameter of ∼100 nm in and high-angle nonequilibrium boundaries with thickness of ∼1 nm. This allows us to relate the obtained materials to nanostructured ones. Currently, SPD is widely used as one of the effective methods for creating bulk nanomaterials with unique properties. First of all, it is a combination of high strength and ductility for various metals and alloys, which is provided by nanostructure.

To obtain submicrocrystalline and nanocrystalline materials, a number of strain accumulation processes are used, as shown in Fig. 9.11.

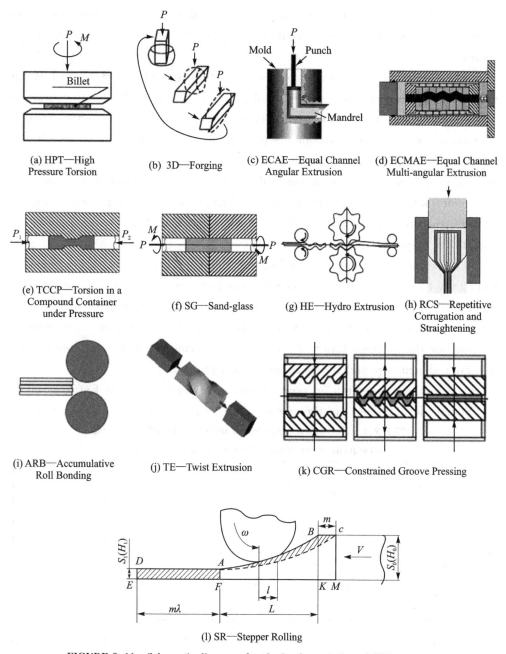

FIGURE 9.11 Schematic diagrams for the implementation of SPD processes

All the methods of severe plastic deformation presented in Fig. 9.11 allow, in varying degrees, to control such important parameters for structural powder products as residual porosity, type of microstructure and size of its components. Eliminating the residual porosity and providing the specified type of microstructure of powder alloys can dramatically improve their performance properties. The effect of severe plastic deformation on the structure and mechanical properties of the powder alloy TC4 were tested using two methods—twist extrusion and stepper rolling.

The essence of screw extrusion method (the method was developed at the Donetsk Institute of Physics and Technology of the National Academy of Sciences of Ukraine) is that a prismatic sample is pressed through a matrix with a channel containing two prismatic sections separated by a section of a screw shape (see Fig. 9.11(j)). During processing, the metal undergoes intense shear deformation. The latter circumstance allows its multiple extrusion to accumulate large deformation, leading to the complete elimination of residual porosity, changes in the structure and properties of the material, which leads to an increase in operational properties.

The essence of the stepper rolling method (see Fig. 9.11(1)) consists in rolling the billet with swaying rolls with simultaneous rotation of the rolled billet around its own axis. This rolling method allows for a single pass to achieve a degree of deformation of 85%, and the presence of a tangential component of deformation due to the rotation of the workpiece, which contributes to obtaining a fine-grained structure and enhanced mechanical properties.

The results of research on severe plastic deformation (Twist Extrusion(TE) and Stepper Rolling(SR)) of the powder alloy TC4 are presented in Table 9.1.

TABLE 9.1 Mechanical properties of the TC4 alloy after sintering, subsequent severe plastic deformation and annealing A

Samples' state	R_m/MPa	A/%	Z/%	KCU/(kJ·m^{-2})
Sintered	950—1 170	10—12	25—27	350—380
Sintered+TE	1 280—1 300	7—8	32—35	350—370
Sintered+TE+A	990—1 100	8—9	39—41	340—350
Sintered+SR	1 380—1 490	4.5—6	8—12	300—320
Sintered+SR+A	1 210—1 300	15—17	52—56	340—380

The transformation of the structure of the sintered TC4 alloy after severe plastic deformation and heat treatment is shown in Fig. 9.12.

From the data presented in Table 9.1 it is shown that severe plastic deformation increases both the strength and its plasticity of the TC4 titanium alloy, eliminates residual porosity, significantly crushes the microstructure, and turns its structure from a plate-like form of α-phase into globular (see Fig. 9.12). Therefore, such methods can be recommended by the industry for the manufacture

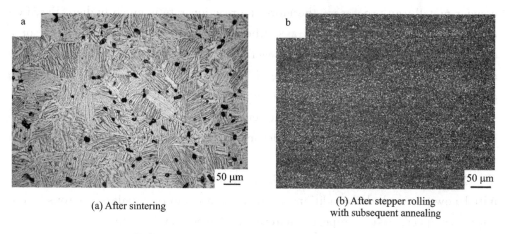

(a) After sintering

(b) After stepper rolling with subsequent annealing

FIGURE 9.12 Microstructure of the TC4 Alloy

of parts with high performance properties.

9.4 Self-propagation High-temperature Synthesis

Self-Propagation High-Temperature Synthesis (SPHTS) of solid chemical compounds is a technological process of obtaining materials based on an exothermic chemical reaction of the interaction of initial reagents in the form of combustion. For organization of the combustion process solid raw materials are used in the form of powders in most cases. In this technology, combustion is not an ordinary reaction of combining with oxygen and the formation of corresponding oxides, but a strongly exothermic reaction of powders with each other or with liquid (gaseous) specially fed reactants (for example, liquefied or gaseous nitrogen) to form solid chemical compounds.

The main principle "not to heat the initial powders, but to burn them" distinguishes the SPHTS technology from the traditional powder metallurgy. Compared with powder metallurgy technology, the SPHTS method has several advantages:

① A small amount of energy to activate the process, that is, the ignition of the original powder. Further, the process proceeds at the expense of its own internal heat release as a result of a strongly exothermic synthesis reaction (self-heating). Energy from the outside is not consumed but is released inside. After synthesis, it must be removed, cooling the products of synthesis.

② Simple and small equipment.

③ High performance. As a result of heating during combustion, very high temperatures are reached, significantly exceeding the heating temperatures in powder metallurgy processes, therefore the reaction rate is much higher. Based on the initial mass of the powder, a synthesis wave is launched in the form of a combustion wave with a speed of several to ten centimetres per second. The duration of the synthesis is from several seconds to several minutes, which is an order of magnitude faster than furnace synthesis.

④ High purity of products and environmental safety, due to very high synthesis temperatures at which decomposition and evaporation of harmful impurities occur.

⑤ A wide range of materials obtained by changing the composition of the initial powders and the conditions of their combustion: powders, porous, non-porous compact, cast, composite materials, surfacing and coatings.

The main disadvantage of the SPHTS method is the necessity for a highly exothermic reaction between the initial powder reactants in order for the product synthesis reaction to take place in the form of combustion so that the initial powder can be ignited.

Since powders of metals and nonmetals are the main reagents in almost all SPHTS processes, and the final product is usually powder material (powder or powder sintered material), the technologies based on SPHTS processes are called *SPHTS powder technologies*, and the resulting materials are *SPHTS powder materials*. The main operations of the SPHTS technology are the preparation of a mixture of powder reagents, the implementation of the SPHTS process itself (combustion of the charge) and the processing of combustion products. SPHTS process has the following distinctive features:

① As reagents, metals (Ti, Ta, Zr, Hf, Mo, and others) usually act as fuel and non-metals (B, C, Si) as oxidizers.

② Depending on the magnitude of the thermal effects of interaction reactions (from 600—900 kJ/kg to 2 900—4 250 kJ/kg, for example, for $Ti+C$ or $Ti+2B$), the level of heat content of the selected system determines the maximum temperature in the combustion zone. It can vary from 800—1 100 ℃ ($Ta+C$, $Ta+B$, $Mo+2Si$) to 2 700—3 400 ℃ ($Hf+C$).

③ To improve the contact between the particles of the reactants, the particle size of the mixture is chosen in the range from 1.0 μm to 200 μm, the density of the pressed charge is 0.1—0.8 (in relative units), and the ratio of reactants is 0.4—2.5 (sometimes the range can be significantly expanded).

④ Depending on the energy content of the reacting system, the initiation of the combustion wave can be carried out either locally on the sample surface (for example, due to glow wire) or by preliminary homogeneous heating of the entire sample in furnaces with a controlled rate of temperature rise up to 600—900 ℃.

⑤ The SPHTS process, after being initiated, is not chaotic. It takes place in a thin layer of the mixture of initial reagents and spreads in the mixture throughout the sample due to heat transfer from hot layers to cold layers of the substance.

The most extensive and studied class of reactions is the SPHTS of gaseous elements. The synthesis reaction of the elements belongs to the category of oxygen-free combustion, where the metal is combustible and the non-metal is an oxidizing agent. However, unlike energy burning, with SPHTS, elements such as C, B, and even H_2 play the role of an oxidizing agent, rather than the fuel.

SPHTSs with participation of chemical compounds as reagents are another extensive class of SPHTS reactions, consisting of four subclasses:

① chemical compounds as an oxidizing agent;
② chemical compounds as flammable;
③ SPHTS with a recovery stage;
④ SPHTS oxides from chemical compounds.

All variants of SPHTS technology are combined into six technological types.

1. SPHTS-powders

SPHTS technology for production of powders is based on the combustion of the initial mixtures of reagents (charges) in special reactors in an inert or reactive gas environment, in vacuum or in air. Combustion products are usually porous materials, which are further processed depending on the type of synthesis (elemental or magnesium thermal) to obtain powders. In case of elemental synthesis, the product is subjected only to mechanical processing (cleaning, grinding, sieving). In case of SPHTS-magnesium-thermium, the product is crushed and subjected to acid enrichment to remove MgO from it and then dried.

2. SPHTS-sintering

With SPHTS-sintering, powder consolidation processes are realized, both in the combustion wave and in the heated (not yet cooled) combustion products. The original mixture of defects is molded in the form of products of a given shape. The burning is organized in such a way that the form is not distorted during the process. The product of combustion is a finished product with a porosity of 5%—50%. Under special conditions, very small, almost zero, and very large (up to 95%) porosity can be achieved.

3. Power SPHTS-compacting

Power SPHTS-compacting consists in compaction of hot SPHTS porous products that did not cool down with external force (pressing, extrusion, rolling, explosion treatment) in order to obtain non-porous materials and products of a given handicap using the plastic properties of combustion products at high temperatures.

4. SPHTS-metallurgy

SPHTS-metallurgy (or SPHTS technology of high-temperature melts) is based on the implementation of liquid-phase SPHTS, in which liquid products are formed during the combustion process. This allows to obtain, when cooled, cast refractory compounds, hard alloys, oxide and composite materials, as well as surfacing from them.

5. SPHTS-welding

With the help of SPHTS-welding, it is possible to obtain a one-piece connection of parts (welding) by conducting this process in the gap between the parts being welded. In this case, the SPHTS process is used as a source of high temperature, and the SPHTS product is used as welding material. Upon reaching the molten state of SPHTS, the product can be removed from the gap (extruded when parts of the workpiece are compressed by externalintervention).

6. Gas transportation SPHTS technology

Gas transportation SPHTS technology is designed for applying thin (5—150 μm) coatings on parts placed in SPHTS-charge. To this end, gas transport additives in the form of halogens or halides are introduced therein. In the wave of combustion, these additives form volatile compounds with the initial reagents. The compounds are transferred to the surface of the part, then break up and form a film of the target product on the surface of the part. Such coatings consist of a film by gas-phase deposition and a wide diffusion zone by diffusion saturation. As a result, they are much more wear-resistant and heat-resistant than the base and have high adhesive strength.

Initially, SPHTS was considered as a new method of obtaining refractory compounds (carbides, borides, nitrides, etc.) due to the use of heat dissipation of a fast exothermic reaction instead of long-term and complex heating of reagents in high-temperature furnaces. However, the synthesis of chemical compounds was the simplest example of the problems solved by the SPHTS method. The method allows to obtain combustion products in the form of final materials with a certain structure and even final products with a given shape and size. It is important that the structural characteristics of combustion products (phase composition, size of crystal grains, distribution of impurities, porosity, etc.) can vary widely depending on the conditions of the SPHTS process and, therefore, can be controlled.

Due to the extreme conditions of the SPHTS process (high temperatures and synthesis rates, self-cleaning effects from impurities, self-annealing and auto-hardening), it is very different from other synthesis methods and allows for fundamentally new possibilities. SPHTS-materials, even having the same chemical composition with known materials, may differ from the latter in structure and properties, and should be treated as new materials.

High quality of SPHTS-powders (purity, dispersion) is also manifested in the final materials, which are usually obtained by hot pressing using activating sintering additives (Y_2O_3, Al_2O_3), as well as in products in which the powders retain their dispersity (backfill, fillers, etc.). For example, ceramics for heat-removing substrates of microcircuits in electronic devices with a very high value $\lambda = 200$ W/(m·℃) was obtained from SPHTS. Dielectric ceramics with tan $\delta = 4.4 \times 10^{-3}$ at frequency of 106 Hz and electrical strength of 9.2 kV/mm are made from SPHTS-powder β-Si_3N_4. On the basis of SPHTS-Si_3N_4 powders with the predominant content of the α-phase, ceramics were obtained with $\rho = 3\,240 - 3\,300$ kg/m^3, porosity less than 1%, $R_{m,\text{ben}} = 850$ MPa, HV=18.0—19.8 GPa.

Based on the structure, the most common are the following three types of SPHTS powders: single crystal, agglomerate and composite.

Single crystal SPHTS powders consist of individual perfect crystals and are obtained as a result of complete grinding of polycrystalline products of combustion. They usually have a grain size of 0.5—3.0 μm, which is easily regulated by the conditions of the SPHTS process. By suppressing recrystallization in combustion products, single crystal powders with grain size of 0.1 μm can be obtained. Coarse-grained single crystal powders are obtained by reducing the cooling rate, which increases the recrystallization time.

Agglomerate SPHTS powders, which cannot be obtained by powder metallurgy, consist of particles formed by individual crystals, between which there is significant splicing. They can be porous. Such powders are formed by incomplete grinding of combustion products. The sizes of particle of agglomerate powders usually range from 10 μm to 200 μm.

Particles of *composite* SPHTS powders consist of phases of various compounds. Minera-ceramic powders containing oxygen-free refractory compounds (carbides, borides) and Al_2O_3 or MgO are widely used. Such powders can be obtained by a single-stage metal-thermal SPHTS process. A typical feature of the powders is a high degree of mixing of the reagents (even inside individual particles). At the same time, they have higher performance properties than powders from mechanical mixtures of the same composition. Composite powder with composition of 90%—70%(Ti, Cr)C+10%—30%Ni is used for protective coatings on parts of aircraft engines.

In recent years, special attention has been paid to nanomaterials and nanotechnologies, and therefore the corresponding direction has been intensively developed in SPHTS technologies. For a long time among the SPHTS powders, the smallest powders were obtained by the azide SPHTS (SPHTS-Az) technology. SPHTS-Az technology is technology for the synthesis of powders of nitrogen-containing compounds (nitrides TiN, ZrN, HfN, AlN, BN, Si_3N_4; carbonitrides, composites Si_3N_4-SiC; Si_3N_4-AlN) in the SPHTS mode, which is based on the use of solid powder as a nitriding agent azide (salts of the acid HN_3) instead of gaseous or liquid nitrogen. Low burning temperatures and the presence of side gaseous products lead to the fact that as a result of the SPHTS-Az process, a loose and very brittle product is obtained. It is very easy to grind by hand into pieces and is also easily grounded to particles less than 40 μm in size.

The first major development in the field of SPHTS was the creation of abrasive pastes based on agglomerated SPHTS powders TiC consisting of separate grains fastened together. Such pastes have a unique ability in one operation to grind and polish the surface of the metal being processed due to the unusual effect of self-destruction of the agglomerate grains during processing. Initially, large, but fairly durable grains carry out "rough work"—grinding the surface. Subsequently, the agglomerate grains are gradually falling apart into small single crystal grains, which begin to carry out "fine work"—polishing the surface (final finishing). Conventional abrasives (Cr_2O_3 and Al_2O_3) do not possess such properties, and titanium carbide, obtained by the kiln or plasma-chemical method, is ineffective when used as abrasive pastes. Abrasive paste is a commercial product consisting of TiC powders and surfactants classified by granularity (fatty acids and their derivatives, hydrocarbons, glycols and their derivatives, polymers and others). The use of titanium carbide pastes instead of diamond abrasives in the processing of non-ferrous metals makes it possible to increase productivity by 1.5—2 times, to increase the surface cleanliness by 1—2 classes, and also to reduce the degree of abrasive caricature into the surface being treated. The most effective use of such pastes was in fine-tuning and polishing parts of aviation equipment, which led to their widespread use in the aviation industry.

Later, SPHTS technology was developed to obtain boron-containing powders (BN, TiB_2, B_4C, $B_{13}C_2$, TiB_2-Al_2O_3, B_4C-Al_2O_3 and others) using the initial reagents of the oxides of the corresponding elements and Mg or Al as a metal reducing agent. Boron-containing SPHTS materials are used as abrasive powders, heat-resistant ceramic products, additives to oils and solid lubricants (BN is used as part of the mixture for the production of sintered friction materials in aircraft).

In recent years, interest in such SPHTS materials as ferroelectrics and ferrites has increased, due to the need to develop new instruments and systems that use the unique properties of such materials for aviation and rocket-space technology.

Ferroelectrics are complex oxide materials used in instrument making, acoustic systems, computing, optical systems and electronics, in the form of various products based on single crystals. Single crystals are grown from ferroelectric powders. Using the SPHTS method, powders of almost the entire range of alkali and alkaline earth metal niobates and tantalates, as well as titanium and zirconate barium and strontium, which are a high-quality mixture for growing single crystals, have been obtained.

Ferrites are chemical compounds of Fe_2O_3 with oxides of other metals (for example, $SrO \cdot 6Fe_2O_3 = SrFe_{12}O_{19}$). Due to the unique combination of magnetic, electrical, dielectric and other properties, ferrites have found wide application in electronics, radio engineering and computing. The main industrial process for the production of ferrite is a long and energy-intensive furnace synthesis. At present, almost all known ferrites are obtained using the SPHTS method. At the same time, the cost of ferrite products based on SPHTS powders is 1.1—1.5 times less, while reducing the spread of electrophysical parameters of products by 2—5 times.

Powders of SPHTS chalcogenides, chemical compounds with S and Se of such metals as Mo, W, Nb, have been obtained. They have a very low coefficient of dry friction and are used as solid lubricants in various dry friction units, as well as anti-friction anti-scuff additives to oils and coolant.

More than 100 binary and complex hydrides have been synthesized by the SPHTS method. The synthesis process is safe, since it proceeds at low hydrogen pressures and is built on its continuous absorption. Combustion temperatures of metals and alloys in hydrogen are relatively not high (200—1 500 ℃). As the initial mixture, the metal in the form of coarse powders, sponges and chips are used. Compared to furnace technology, a higher hydrogen content in hydrides has been achieved (it is possible to obtain superstoichiometric compounds). SPHTS hydrides are intended for use as condensed hydrogen carriers, efficient hydrogen accumulators and, in the long term, components of environmentally friendly hydrogen fuel. Currently, hydrides are used as feedstock for the production of powder structural alloys, in particular titanium, of improved quality, since the atomic hydrogen released during the initial stage of sintering products is an effective reducing agent that removes harmful impurities from the charge.

SPHTS products, as a rule, have high porosity. Their porous structure depends on a number of factors. The main ones are the composition and structure of the charge, the volumetric rate of release of impurity gases and the presence of liquid products in the combustion wave. By varying these parameters, it is possible to adjust the porosity of SPHTS products in a wide range depending on the application purpose (usually 40%—70% porosity).

Porous SPHTS materials have a number of advantages in comparison with the same materials synthesized by powder metallurgy methods. First of all, there exists a very high final porosity, due to the negative volume effect of the SPHTS reaction and the release of impurity gases during the conversion of the charge into the final product (with the usual sintering of ceramic materials, no such effects are observed). High porosity helps to reduce energy consumption with further grinding of the product for final use as a powder.

Another advantage of SPHTS materials is that their strength is 1.5—3 times higher than that of sintered materials with the same porosity. The high temperatures of the SPHTS process and the low content of impurities at the grain boundaries contribute to the formation of strong bonds between the grains in the polycrystal, which is a obvious advantage of this method, especially when creating composite materials.

The high strength of the SPHTS porous TiC framework with subsequent impregnation with its standard superalloy nickel alloy allowed us to obtain a composite material with heat resistance properties substantially superior to the standard nickel alloy. Porous frameworks made of TiC by sintering in a furnace were destroyed when impregnated with a nickel alloy, and the effect of improving heat resistance was not observed.

One of the options for industrial use of SPHTS technology is the production of porous filter components from corrosion and heat resistant refractory compounds (carbides, nitrides, oxides), in particular for filtering fuels, oils and hydraulic fluids for aviation equipment. Due to the capabilities of SPHTS technology, filters can have both a uniform and a gradient structure.

Highly porous SPHTS materials (with porosity of more than 90%) are obtained by adding additives ($Na_2B_4O_7 \cdot 10H_2O$, TiH_2, Ca_2F_4, NH_4Cl) to the mixture that is gasified during the SPHTS process, as well as during the SPHTS process under conditions of microgravity (weightlessness). At the same time, highly porous materials can be formed, both predominantly open and predominantly closed porosity (foams).

One of the options for obtaining ceramic materials with high properties and density is gas-static SPHTS technology, in which the synthesis process is combined with a high pressure of gaseous nitrogen, which allows you to directly synthesize following products:

① structural nitride and nitride-carbide ceramics without activating sintering additives;

② heat-resistant and corrosion-resistant ceramics based on sialons, aluminum nitride and its compositions with transition metal borides;

③ functional ceramics based on Si_3N_4 and AlN;

④ tribotechnical ceramics based on Si_3N_4, SiC and BN;

⑤ new multi-component compositions of non-metallic nitrides, silicon carbides, aluminum, boron with refractory metal-like compounds.

The ceramics obtained using SPHTS process is distinguished by abnormally high corrosion resistance in metal melts, increased resistance to thermal shocks, and high hardness. SPHTS ceramics is able to retain their properties in a wide range of temperatures, which is a huge advantage over traditional ceramics. This is explained by the poor sintering of ceramic powders, and therefore some oxides (MgO, Y_2O_3 and others) are additionally introduced to promote sintering. At high temperatures, such additives melt, which leads to a decrease in the strength of traditional sintered ceramics. SPHTS ceramics does not contain such oxide additives and is free from this drawback.

Using SPHTS technology, Functional Gradient Materials (FGM) are produced. In FGM, the local composition, and, consequently, the properties, vary in the volume of the sample in accordance with a given pattern. Sometimes they are called materials with pre-planned heterogeneity. Since the properties of materials often non-linearly depend on the composition of the material, the general properties of the FGM can differ significantly from homogeneous material of the same average composition, which allows you to adjust the properties of products and produce products with unusual properties.

To obtain FGM using the SPHTS method of compaction, the initial charge billet consists of two or more layers of different chemical compositions. In the two-layer scheme, the first layer is the reaction mixture, the second layer is inert, consisting of pure metal or alloy. SPHTS compaction is one of the most efficient methods for producing gradient products of various sizes, and the products themselves can be successfully used under the combined effect of abrasive and impact loads.

Impact-resistant TiC + Ni carbide plates are obtained by SPHTS pressing of blanks of a gradient material and subsequent machining of these blanks into plates of the required size. The hardness of the plate thickness varies from 82 HRA to 90 HRA, and flexural strength is from 1 000 MPa to 1 200 MPa. The impact strength of plates made of gradient material is 3 times higher than the impact strength of standard hard alloy plates. Gradient plates are designed to work under shock loads and increased wear (protecting helicopters from being hit by bullets and shrapnel).

The SPHTS technology has been developed for producing of targets for magnetron and ion-plasma sputtering installations for resistive, wear-resistant and corrosion-resistant coatings on parts of aviation and rocket-space technology. Targets can be multi-component, multi-phase and functionally gradient. They have a high density, uniform structure, high level of mechanical, thermal and electrical properties.

The target materials are non-stoichiometric carbides (TiC_x, $(Ti, Cr)C_x$, $(Ti, Nb)C_x$, $(Ti, Mo)C_x$); borides (TiB, ZrB, CrB); silicides ($TiSi_2$, Ti_5Si_2, $MoSi_2$, Cr_5Si_3, Ni_5Si_2); intermetallic compounds (Ti-Al, Ni-Al, Ti-Al-Si, Ti-Nb-Al, Ti-Zr-Al); chalcogenides ($MoSe_2$, WSe_2, MoS_2); composites ($TiB_2 + Ti_5Si_3 + Si$, $TiC + Ti_3SiC + TiSi_2$, $TiC + Ti_3SiC_2 + SiC$, $TiC + TiAl$, $TiB_2 + TiC$, $TiB_2 + TiAl + Ti_2AlN$). Such materials are difficult to deform and sinter, so targets from them are difficult, and sometimes impossible, to be obtained using traditional powder metallurgy technology. The heat resistance and service life of SPHTS targets are 3—5 times higher than those produced by traditional methods.

The new direction of improving the targets for magnetron sputtering is introduction of metal nanopowders or refractory compounds into their composition. Such additives modify the structure of wear-resistant coatings and significantly improve their physical and mechanical properties (density, microhardness, crack resistance).

Using SPHTS-casting method, ingots of refractory compounds such as carbides, silicides, intermetallics, oxides, as well as alloys and composites based on them are obtained. Ingots of these materials are easily crushed into powders of various fractions. Carbide powders obtained by the SPHTS method have several advantages when used in aircraft construction as a material for applying plasma wear-resistant high-temperature coatings. In traditional plasma spraying technology, Cr_3C_2 or $TiC-Cr_3C_2$ powder is used, which is additionally plated with nickel. The developed technology of single-stage SPHTS-casting of $TiC-Cr_3C_2-Ni$ composite eliminates the operation of nickel carbide cladding in special equipment. Plasma coatings from such composite powders are more than 10 times better than clad ones in wear resistance. In addition, composite coatings have a sufficiently high heat resistance and withstand more than 1 000 heat cycles in the temperature range from 700 ℃ to 100 ℃ and 200 heat cycles in the temperature range from 900 ℃ to 100 ℃ without cracks or delaminations.

The creation of new high-strength materials with improved properties, capable of operating in extreme conditions (high temperature, chemically aggressive environment, significant mechanical loads, and so on) is the limiting stage in the development of modern aviation and rocket-space technology. The remarkable balance of properties of titanium aluminides, including thermodynamic stability, high heat resistance combined with low density, determine the prospects of their use in aircraft and rocket production.

Ni_3Al and NiAl are obtained by SPHTS-casting. From NiAl the specialists get a powder for producing protective coatings (in particular surfacing). Composite powders of the compositions $Cr_3C_2+Ni-Al$, $(Cr-Ti)B+Ni-Al-Mn$ are solid grains of chromium and titanium carbides (borides) distributed in a Ni-Al-(Mn) matrix. Such matrix has a melting point close to that of a steel base. Therefore, during coating, a melt is created in which the carbide (boride) grains do not melt and do not have time to dissolve in the matrix. As a result, the coating has a high hardness and wear resistance, which exceeds the surfacing of industrial cast alloy "Sormit" in 2—4 times. Welding electrodes from SPHTS-Ni_3Al casting are used for welding technological holes and defects in GTE blades from industrial nickel superalloy using argon-arc welding.

The FeAl cast exceeds the stainless austenitic nickel-chrome steel in heat resistance in 45 times and is promising for use as heat-resistant coatings. Cast Nb_3Al has a transition temperature to the superconducting state of -254.4 ℃ and is of interest for the preparation of superconductors.

Liquid-phase SPHTS allows to obtain multicomponent cast oxide materials based on Al_2O_3 and Cr_2O_3. Such materials possess not only high abrasive properties, but also unique chemical resistance and heat resistance. The Al_2O_3-Cr_2O_3 system has unlimited mutual solubility of oxides. SPHTS technology allows to obtain cast oxides in this system with a Cr_2O_3 content from 6% to 100%. Solid oxide solutions based on Al_2O_3 and Cr_2O_3 are promising for manufacturing of casting molds, the use of which significantly improves the quality of the blades, thereby increasing the resource of gas turbine engines. To summary, SPHTS technology is very promising for aircraft and rocket production.

9.5 Metal Injection Molding

Metal Injection Molding (*MIM*) is a technological process for manufacturing small precision parts (including thin-walled multi-stage in height) with a fine-grained structure without defects and anisotropy of mechanical properties that are difficult, very expensive, or impossible to manufacture using traditional methods. MIM technology has been one of the three fastest growing technologies in powder metallurgy in the past two decades (together with HIP and the application of additive manufacturing)[184].

The design of parts manufactured by powder metallurgy and the field of their application develop in several directions. One of them is the complication of the form, and the second is miniaturization. The use of MIM technology makes it possible to expand the use of powder parts in products where very small high-precision parts are needed, for example, for instrument-making and electronics, without which modern aviation and rocket-space technology is impossible.

With the help of MIM technology, it is possible to create complex forms from almost all materials, including metals, ceramics, intermetallics, composite materials, cermet (TiC-steel, TiN-steel, WC-Co, TiC-Ni), as well as carbon, high-speed and corrosion-resistant alloys based on Al, Cu, Ti, Fe-Ni, Fe-Ni-Co, etc.

Metal injection molding is a development of traditional powder metallurgy technology. When manufacturing parts using standard powder metallurgy technology, there is one significant limitation in their configuration. So, after sealing the part in the mold, the part must be removed, i. e. pushed out of the mold cavity. It is obvious that parts with undercuts (grooves) or protrusions located at right angles (or obliquely) to the direction of extrusion cannot be made directly according to the traditional powder metallurgy scheme. Due to MIM technology, restrictions on configuration are largely eliminated.

The main technological operations of the MIM process are mixing the powder with the binder, injecting the mixture into the working cavity of the matrix, removing the part from the matrix, removing the binder and sintering.

Almost any metal that can be processed into appropriate powders can be used in MIM technology (with the exception of Al and Mg due to the presence of an oxide film on the surface of the particles, which makes it difficult to sinter, and also that Ti is embrittled during sintering by a binder). From an economic point of view, the most promising candidates are expensive materials since MIM technology has virtually no waste, which compensates the high cost of the powder.

Basic requirements for powders used in MIM technology are as follows:

① given particle size distribution to ensure high packing density;

② lack of agglomeration;

③ predominance of spherical (or equiaxial) particle shape;
④ interparticle friction;
⑤ sufficient to avoid deformation after removing the binder;
⑥ small particle size (less than 20 μm) for fast sintering;
⑦ dense particles without internal voids;
⑧ minimal explosiveness and toxicity;
⑨ clean particle surface for predictable interaction with the binder.

The methods used in powder metallurgy to study the properties of powders (screen analysis, determination of flowability or compressibility) are not applicable to MIM powders, since their particle size is much smaller than that of powders intended for pressing in a mold. In the case of MIM powders, specific surface areas are monitored (ISO10070) and gravity sedimentation is considered (ISO10076).

Binder is an important factor for the successful production of MIM parts. In most cases, binders are mixtures of organic compounds, whose main components are natural wax or artificial polymers, whose properties can be modified by adding other substances.

Basic requirements for binders include:
① lack of chemical interaction with the metal;
② no deterioration during the process;
③ lightness and completeness of removal from the formed part.

Mixing of powders with a binder is carried out in Z-blade and planetary mixers, as well as in twin-screw extruders. When mixing, the main task is to ensure that the entire surface of the powder particles is coated with a binder. Therefore, mixing is carried out at elevated temperatures, when the binder becomes fluid and "wets" the powder particles, which ensures the formation of a homogeneous mixture without clusters of particles. To facilitate and enhance the contact between the surface of the particle and the binder, sometimes the powder is pretreated. In addition, surfactants are often added to the mixture.

In the production of MIM parts it is necessary to use as little binder as possible. The volume fraction of the powder depends on its characteristics and varies, as a rule, in the range from 0.5 to 0.7.

The rheological properties of raw materials—a mixture of powder and binder play an important role in MIM technology. The viscosity of the raw material at the injection temperature must be such as to ensure a smooth flow of the mixture into the matrix without any segregations. In addition, the viscosity of the mixture should be as constant as possible in a given temperature range. At the same time the mixture should harden when cooled[176].

The resulting mixture of powder with a binder is processed into solid granules that can be stored and, if necessary, fed into a molding machine. Injection molding machines are similar to those in the plastics industry. The screw with which the mixture is extruded into the cavity of the matrix, is located in a heated cylinder. To ensure constant process conditions, the temperature of the cylinder and nozzle is carefully controlled. The temperature of the matrix is also controlled: it must be low enough so that the molding remains "solid" in the process of moving. The machines are equipped with a self-adjusting controller that allows to adjust the speed and pressure of injection, pressing pressure and other parameters.

In addition to the standard manufacturing process, which is called high pressure injection molding, medium and low pressure injection molding is also used. The advantage of lower injection pressures is lower investment costs for equipment and tooling.

Removing the binder from the "Green" part is an important technological operation and requires careful monitoring. There are several basic ways to remove a binder:

① Heating "Green" parts to melt, decompose and, ultimately, evaporate the polymer binder. The process is carried out with good control in order to avoid the destruction of the part. It is advisable to use a binder with several components that decompose or evaporate at different temperatures. This process usually lasts several hours, depending on the wall thickness of the part.

② Catalytic decomposition of polyacetate MIM raw materials using gaseous nitric or oxalic acids. This significantly reduces the time to remove the binder and the probability of destruction of the part. There has been developed such equipment that allows continuous catalytic binder removal and sintering.

③ Dissolving the binder with appropriate solvents such as acetone, ethyl alcohol or hexane. Some components of the binder may even be water soluble. Heating is usually the final operation to complete the removal of the binder by evaporation.

④ Removal of the binder using supercritical CO_2 as a solvent. The supercritical state of CO_2 is achieved at elevated pressures and temperatures and is intermediate between gaseous and liquid states. It is characterized by a very low viscosity of the solvent, which reduces the time for removal of the binder.

Sintering in MIM technology is essentially the same process as in traditional powder metallurgy. As a rule, reducing environments are used to prevent oxidation of the metal and reduction of oxides on the surface of the powder particles. In MIM technology, cleaner powders are used, which is more important for MIM technology than for traditional powder metallurgy. The increased purity of MIM powders causes an easier sintering process due to the higher surface energy of the particles. In traditional powder metallurgy technology, sintered parts have dimensions that are usually very close to the size of the original pressed billet. Therefore, with this technology, it is not difficult to provide tight dimensional tolerances. In case of injection molding, the situation is different. Since the "Green" part contains a large amount of binder (50%), a considerable shrinkage occurs in the sintering process, which requires very tight control. In this respect, the advantage of MIM technology compared with traditional powder metallurgy is that the density of the metal in the molding is uniform, provided that the mixture is prepared correctly. Consequently, the shrinkage during sintering of such a part, though large, is also uniform. This eliminates the possibility of deformation of the part during sintering, which may occur due to the uneven density of the part pressed in the mold.

Finished MIM is the part with density close to theoretical. The mechanical properties of MIM parts are similar to those for deformed metal of similar composition. The properties of MIM parts can be improved with the help of standard technological operations used for deformable metals and/or parts of powder metallurgy, e.g., cementation, electroplating. Often, the surface of MIM is so dense that the residual internal porosity does not adversely affect subsequent technological operations.

Theoretically, there are no restrictions on the maximum size of MIM parts, but from an economic point of view, their size is still limited. In this consideration, two points should be noted:

① For large parts, their total cost is greater, due to the use of more expensive raw materials. The total cost of the powders is a linear function of the mass of the part. However, in case of parts manufacturing, for example, by machining from a rod or other rational billet, the increase in the cost of machining with the increase of the part size occurs not so frequently.

② With the increase in the size of the cross section, the time for removing the binder also increases and, consequently, the cost of this technological operation increases. Therefore, at present, the thickness is limited to ~30 mm.

Thus, injection molding, in fact, is a technology for manufacturing parts of complex shapes and small sizes. If the shape of the part allows it to be made, for example, by ordinary pressing and sintering (and the mechanical properties are satisfactory), the use of MIM technology is in most cases impractical due to the high cost. However, if the required number of complex parts is more than a certain amount, the use of MIM technology becomes cheaper than mechanical processing.

Precision casting typically competes with MIM technology. In many characteristics, MIM technology has advantages over precision casting. MIM technology has advantages over precision casting in case of mass production of small parts and alloys with complex compositions. Fig. 9.13 gives examples of variously structured parts produced by MIM technology.

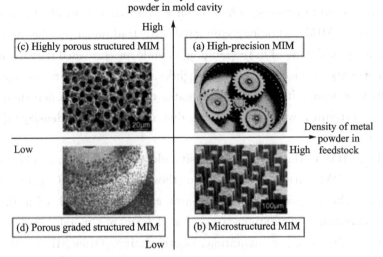

FIGURE 9.13 Variously structured MIM parts determined by homogeneity of metal powder in mold cavity and density of metal powder in feedstock[185]

However, MIM technology has some drawbacks, including the relatively high cost of fine powders, the need for strict adherence to exact technological processing conditions, the complexity of the processes of removing the binder, and expensive specialized equipment. Despite this, MIM technology will develop at a rather rapid pace, as modern technology (including aviation and rocket-space technology) requires an increasing number of accurate small-sized parts from complexly alloyed materials with predictable special properties.

References

[1] Giurgiutiu V. Structural Health Monitoring of Aerospace Composites. Oxford: Academic Press, 2016.

[2] Krivov G A. Technology Aircraft Manufacturing Production [In Russian]. Kyiv: KVITS Publ, 1997.

[3] Koster J, Velazco A, Munz C D, et al. HYPERION UAV: An International Collaboration, 50th AIAA Aerospace Sciences Meeting Including the New Horizons Forum and Aerospace Exposition. Tennessee: AIAA, 2012:1223.

[4] Njuguna J. Lightweight Composite Structures in Transport: Design, Manufacturing, Analysis and Performance. Cambridge: Woodhead Publishing, 2016.

[5] Molyar A, Kotsyuba A, Bychkov A, et al. Structural Materials in Aircraft Manufacture [In Russian]. Kyiv: KVIC, 2015.

[6] Williams J C, Starke E A. Progress in Structural Materials for Aerospace Systems. Acta Materialia, 2003, 51: 5775-5799.

[7] Zhang X, Chen Y, Hu J. Recent Advances in the Development of Aerospace Materials. Progress in Aerospace Sciences, 2018, 97: 22-34.

[8] Prasad N E, Wanhill R J. Aerospace Materials and Material Technologies. Berlin: Springer, 2017.

[9] Watanabe R, Kawasaki A. Development of Functionally Gradient Materials via Powder Metallurgy. Journal of the Japan Society of Powder and Powder Metallurgy, 1992, 39: 279-286.

[10] Zhu J, Lai Z, Yin Z, et al. Fabrication of Zro2-Nicr Functionally Graded Material by Powder Metallurgy. Materials Chemistry and Physics, 2001, 68: 130-135.

[11] Erdemir F, Canakci A, Varol T, Microstructural Characterization and Mechanical Properties of Functionally Graded Al2024/Sic Composites Prepared by Powder Metallurgy Techniques. Transactions of Nonferrous Metals Society of China, 2015, 25: 3569-3577.

[12] Battabyal M, Schäublin R, Spätig P, et al. W-2 wt.% Y_2O_3 Composite: Microstructure and Mechanical Properties. Materials Science and Engineering A, 2012, 538: 53-57.

[13] Tsuge A, Nishida K, Komatsu M. Effect of Crystallizing the Grain-Boundary Glass Phase on the High-Temperature Strength of Hot-Pressed Si_3N_4

Containing Y_2O_3. Journal of the American Ceramic Society, 1975, 58: 323-326.

[14] Takagi K. Development and Application of High Strength Ternary Boride Base Cermets. Journal of Solid State Chemistry, 2006, 179: 2809-2818.

[15] Semchysen M. Refractory Metals and Alloys//Bever M B. Encyclopedia of Materials Science and Engineering. Oxford: Pergamon Press, 1986.

[16] Burkanov G S, Efimov Y V. Refractory Metals and Alloys [In Russian]. Moscow: Metallurgiya, 1986.

[17] Raymond O, Perera L C, Brothers P J, et al. The Chemistry and Metallurgy of Beryllium. Chemistry in New Zealand, 2015, 79: 137-143.

[18] Naik B, Sivasubramanian N. Applications of Beryllium and Its Alloys. Mineral Processing and Extractive Metullargy Review, 1994, 13: 243-251.

[19] Kostornov A, Fushchich O. Sintered Antifriction Materials. Powder Metallurgy and Metal Ceramics, 2007, 46: 503-512.

[20] Fedorchenko I M, Pugina L I. Composite Sintered Antifriction Materials. Kiev: Ndumka, 1980.

[21] Yao P P, Sheng H C, Xiong X, et al. Worn Surface Characteristics of Cu-Based Powder Metallurgy Bake Materials for Aircraft. Transactions of Nonferrous Metals Society of China, 2007, 17: 99-103.

[22] Xiao P, Xiong X, Zhang H B, et al. Progress and Application of C/C-Sic Ceramic Braking Materials. Chinese Journal of Nonferrous Metals, 2005, 15: 667-674.

[23] Suzdaltsev E. Radio Transparent, Heat-Resistant Materials for the 21st Century. Refractories and Industrial Ceramics, 2002, 43: 103-110.

[24] Choi J, Ishitsuka H, Mima S, et al. Radio-transparent Multi-Layer Insulation for Radiowave Receivers. Review of Scientific Instruments, 2013, 84: 114502.

[25] Orlova L, Chainikova A, Alekseeva L, et al. Recent Advances in Radio Transparent Glass-Ceramic Materials Based on High-Temperature Aluminosilicate Systems. Russian Journal of Inorganic Chemistry, 2015, 60: 1692-1707.

[26] Han L, Song J, Zhang Q, et al. Crystallization, Structure and Characterization of $Mgo-Al_2O_3-Sio_2-P_2O_5$ Transparent Glass-Ceramics with High Crystallinity. Journal of Non-Crystalline Solids, 2018, 481:123-131.

[27] Hou X, Zhou S, Jia T, et al. White Light Emission in $Tm^{3+}/Er^{3+}/Yb^{3+}$ Tri-Doped Y_2O_3 Transparent Ceramic. Journal of Alloys and Compounds, 2013, 509: 2793-2796.

[28] Heydari M S, Ghezavati J, Abbasgholipour M, et al. Various Types of Ceramics Used in Radome: A Review. Scientia Iranica Transaction B: Mechanical Engineering, 2017, 24: 1136-1147.

[29] Dao M, Lu L, Asaro R, et al. Toward a Quantitative Understanding of Mechanical Behavior of Nanocrystalline Metals. Acta Materialia, 2007, 55: 4041-4065.

[30] Kumar K, van Swygenhoven H, Suresh S. Mechanical Behavior of Nanocrystalline Metals and Alloys. Acta Materialia, 2003, 51: 5743-5774.

[31] Ramesh K T. Nanomaterials: Mechanics and Mechanisms. Berlin: Springer, 2009.

[32] Lowe T C, Zhu Y T. Commercialization of Nanostructured Metals Produced by Severe Plastic Deformation Processing. Advanced Engineering Materials, 2003, 5: 373-378.

[33] Sorensen K L, Helland A S, Johansen T A. Carbon Nanomaterial-Based Wing Temperature Control System for In-Flight Anti-Icing and De-Icing of Unmanned Aerial Vehicles. 2015 IEEE Aerospace Conference, IEEE, 2015: 1-6.

[34] Micheli D, Vricella A, Pastore R, et al. Synthesis and Electromagnetic Characterization of Frequency Selective Radar Absorbing Materials Using Carbon Nanopowders. Carbon, 2014, 77: 756-774.

[35] Dhakate S R, Subhedar K M, Singh B P. Polymer Nanocomposite Foam Filled with Carbon Nanomaterials as an Efficient Electromagnetic Interference Shielding Material. Rsc Advances, 2015, 5: 43036-43057.

[36] Liu R, Wang Z, Sparks T, et al. Aerospace Applications of Laser Additive Manufacturing. Laser Additive Manufacturing, Elsevier, 2017.

[37] Antonysamy A A. Microstructure, Texture and Mechanical Property Evolution During Additive Manufacturing of Ti6Al4V Alloy for Aerospace Applications. Manchester: The University of Manchester, 2012.

[38] Kamal M, Rizza G. Design for Metal Additive Manufacturing for Aerospace Applications. Additive Manufacturing for the Aerospace Industry, Elsevier, 2019.

[39] Uriondo A, Esperon-Miguez M, Perinpanayagam S. The Present and Future of Additive Manufacturing in the Aerospace Sector: A Review of Important Aspects. Proceedings of the Institution of Mechanical Engineers: Part G Journal of Aerospace Engineering, 2015, 229: 2132-2147.

[40] Grzesiak A, Becker R, Verl A. The Bionic Handling Assistant: A Success Story of Additive Manufacturing. Assembly Automation, 2011, 31: 329-333.

[41] Subrahmanyam J, Vijayakumar M. Self-Propagating High-Temperature Synthesis. Journal of Materials Science, 1992, 27: 6249-6273.

[42] Kurbatkina V, Patsera E, Levashov E, et al. Self-Propagating High-Temperature Synthesis of Refractory Boride Ceramics (Zr, Ta) B_2 with Superior Properties. Journal of the European Ceramic Society, 2018, 38: 1118-1127.

[43] Libenson G. Foundations of Powder Metallurgy [In Russian]. USSR: INIS, 1987.

[44] Tracey V. Sintering of Porous Nickel. Powder Metallurgy, 1983, 26: 89-92.

[45] Iacocca R, German R. A Comparison of Powder Particle Size Measuring Instruments. International Journal of Powder Metallurgy, 1997, 33: 35-48.

[46] Iacocca R G. Particle Size and Size Distribution//Eisen W B, Ferguson B L, German R M, et al. ASM Handbook: Vol 7 Powder Metallurgy Technologies and Applications, Materials Park. OH: ASM International, 1998: 234-238.

[47] Webb P A. Volume and Density Determinations for Particle Technologists. Micromeritics Instrument Corp, 2001, 2: 1-16.

[48] Mathews D. Measurement of the Specific Surface Area of Fine Powders: A Comparison of the "Gas-adsorption" and "Air-permeability" Methods. Journal of Applied Chemistry, 1957, 7: 610-613.

[49] Naderi M. Surface Area: Brunauer-Emmett-Teller (BET)//Tarleton S. Progress in Filtration and Separation. Cambridge: Academic Press, 2015: 585-608.

[50] Stanley-Wood N G. Enlargement and Compaction of Particulate Solids. Oxford: Butterworth-Heinemann, 1983.

[51] Bogatyreva G, Petasyuk G, Bazalii G, et al. On Morphometric Uniformity of Diamond Micron Powders. Journal of Superhard Materials, 2009, 31: 126-134.

[52] Angelo P, Subramanian R. Powder Metallurgy: Science, Technology and Applications. New Delhi: PHI Learning, 2008.

[53] Narasimhan K S. Powder Characterization//Buschow K H J, Cahn R W, Flemings M C, et al. Encyclopedia of Materials: Science and Technology. Oxford: Elsevier, 2001:7781-7788.

[54] Beakawi Al-Hashemi H M, Baghabra Al-Amoudi O S. A Review on the Angle of Repose of Granular Materials. Powder Technology, 2018, 330: 397-417.

[55] Krantz M, Zhang H, Zhu J. Characterization of Powder Flow: Static and

Dynamic Testing. Powder Technology, 2009, 194: 239-245.

[56] Karataş C, Sözen A, Arcaklioglu E, et al. Investigation of Mouldability for Feedstocks Used Powder Injection Moulding. Materials & Design, 2008, 29: 1713-1724.

[57] Samal P, Newkirk J. Compressibility and Compactibility of Metal Powders. Almere: ASM International, 2015.

[58] Angelo P C. Subramanian R Powder Metallurgy: Science, Technology and Applications. New Delhi: PHI Learning, 2008.

[59] Antsiferov V, Bobrov G, Druzhinin L. Powder Metallurgy and Sprayed Coatings. Moscow: Publ Metallurgiya, 1987.

[60] Rakovskii V S. Principles of Physical Powder Metallurgy: Translation Division, Foreign Technology Division. Generic, 2018.

[61] Krizik P, Balog M, Bajana O, et al. Warm Pressing of Al Powders: An Alternative Consolidation Approach//Ratvik A P. Light Metals. Cham: Springer International Publishing, 2017: 463-469.

[62] Atkinson H, Davies S. Fundamental Aspects of Hot Isostatic Pressing: An Overview. Metallurgical and Materials Transactions A, 2000, 31: 2981-3000.

[63] Helle A, Easterling K E, Ashby M. Hot-Isostatic Pressing Diagrams: New Developments. Acta Metallurgica, 1985, 33: 2163-2174.

[64] Grigorev E, Kalin B. Electropulse Technology for the Formation of Materials Made of Powders [In Russian]. Moscow: MIFI, 2008.

[65] Danninger H, Gierl-Mayer C. Advanced Powder Metallurgy Steel Alloys//Chang I, Zhao Y. Advances in Powder Metallurgy. Cambridge: Woodhead Publishing, 2013:149-201.

[66] Gogaev K, Kalutskii G Y, Voropaev V. Asymmetric Rolling of Metal Powders: I Compactability of Metal Powders in Asymmetric Rolling. Powder Metallurgy and Metal Ceramics, 2009, 48: 152-156.

[67] German R M. Liquid Phase Sintering. Berlin: Springer, 2013.

[68] Hryha E, Nyborg L, Malas A, et al. Carbon Control in PM Sintering: Industrial Applications and Experience. London: Taylor & Francis, 2013.

[69] Upadhyaya G S. Powder Metallurgy Technology. Cambridge: Cambridge Int Science Publishing, 1997.

[70] Binns C. Introduction to Nanoscience and Nanotechnology. Hoboken, NJ: John Wiley & Sons, 2010.

[71] Iqbal P, Preece J A, Mendes P M. Nanotechnology: The "Top-Down" and "Bottom-Up" Approaches. Supramolecular Chemistry: From Molecules to Nanomaterials, 2012: 1-14.

[72] Shimomura M, Sawadaishi T. Bottom-Up Strategy of Materials Fabrication: A New Trend in Nanotechnology of Soft Materials. Current Opinion in Colloid & Interface Science, 2001, 6: 11-16.

[73] Wardle B L, Cohen E, Cornwell H, et al. Aligned CNT-Based Microstructures and Nanoengineered Composite Macrostructures. Nanotechnology and Nanomaterials, 2017, 125.

[74] Pokropivny V, Skorokhod V. Classification of Nanostructures by Dimensionality and Concept of Surface Forms Engineering in Nanomaterial Science. Materials Science and Engineering C, 2007, 27: 990-993.

[75] Mourdikoudis S, Pallares R M, Thanh N T. Characterization Techniques for Nanoparticles: Comparison and Complementarity upon Studying Nanoparticle Properties. Nanoscale, 2018, 10: 12871-12934.

[76] Matrenin S, Ovechkin B. Nanostructured Materials in Machine Engineering [In Russian]. Tomsk: Politekhn. Univ, 2009.

[77] Poole C P, Owens F J. Introduction to Nanotechnology, Hoboken, NJ: John Wiley & Sons, 2003.

[78] Hasanovo. Methods of Compacting and Consolidating Nanostructured Materials and Products. Tomsk: Tomsk Polytechnic University Publ, 2008.

[79] Kaygorodov A S, Ivanov V V, Khrustov V R, et al. Fabrication of Nd: Y_2O_3 Transparent Ceramics by Pulsed Compaction and Sintering of Weakly Agglomerated Nanopowders. Journal of the European Ceramic Society, 2007, 27: 1165-1169.

[80] Munir Z, Anselmi-Tamburini U, Ohyanagi M. The Effect of Electric Field and Pressure on the Synthesis and Consolidation of Materials: A Review of the Spark Plasma Sintering Method. Journal of Materials Science, 2006, 41: 763-777.

[81] Ragulyaa V. Fundamentals of Spark Plasma Sintering//Buschow K H J, Cahn R W, Flemings M C, et al. Encyclopedia of Materials: Science and Technology. Oxford: Elsevier, 2010:1-5.

[82] Gusev A. Nanomaterials, Nanostructures, and Nanotechnologies. 2nd ed. Moscow: Nauka-Fizmatlit, 2007.

[83] Berbentsev V, Alymov M, Bedov S. Nanopowder Consolidation by Gas Extrusion Method. Nanotechnologies in Russia, 2007, 7-8: 116-120.

[84] Dursun T, Soutis C. Recent Developments in Advanced Aircraft Aluminium Alloys. Materials & Design, 2014, 56: 862-871.

[85] Fridlyander I N. Aluminum Alloys [In Russian]. Kiev: Kominteh, 2005.

[86] Huo S, Qian M, Schaffer G, et al. Fundamentals of Aluminium Metallurgy. Cambridge: Woodhead Publishing, 2011: 655-701.

[87] Belov A, Anoshkin N, Khodkin V. Granule Metallurgy—A New, Advanced Technological Process for the Manufacture of Materials//Processing of Light and Heat-Resisting Alloys [In Russian]. Moscow: Nauka, 1976: 217-236.

[88] Kharitonov A, Sheikhaliev S M. Production of Metal Powders from Melts by Centrifugal Atomization. Soviet Powder Metallurgy and Metal Ceramics, 1985, 24: 883-887.

[89] Chen C, Guo L, Luo J, et al. Aluminum Powder Size and Microstructure Effects on Properties of Boron Nitride Reinforced Aluminum Matrix Composites Fabricated by Semi-Solid Powder Metallurgy. Materials Science and Engineering A, 2015, 646: 306-314.

[90] Radomyselskii I, Serdyuk G, NI S. Structural Powder Materials. Kiev: Tekhnika, 1985.

[91] Fridlyander I, Chuistov K, Berezina A, et al. Aluminum-Lithium Alloys: Structure and Properties. Kiev: Naukova Dumka, 1992.

[92] Ilyin A, Kolachev B, Polkin I. Titanium Alloys: Composition, Structure, Properties [In Russian]. Moscow: VILS-MATI, 2009.

[93] Qian M, Froes F H. Titanium Powder Metallurgy: Science, Technology and Applications. Oxford: Butterworth-Heinemann, 2015.

[94] Liu Y, Chen L, Tang H, et al. Design of Powder Metallurgy Titanium Alloys and Composites. Materials Science and Engineering A, 2006, 418: 25-35.

[95] Yolton C. The Pre-Alloyed Powder Metallurgy of Titanium with Boron and Carbon Additions. JOM, 2004, 56: 56-59.

[96] Froes F, Eylon D, Eichelman G, et al. Developments in Titanium Powder Metallurgy. JOM, 1980, 32: 47-54.

[97] Shanmugam S. Granulation Techniques and Technologies: Recent Progresses. Bioimpacts, 2015, 5: 55-63.

[98] Ivasishin O, Sawakin D, Moxson V, et al. Titanium Powder Metallurgy for Automotive Components. Materials Technology, 2002, 17: 20-25.

[99] Fang Z Z, Paramore J D, Sun P, et al. Powder Metallurgy of Titanium—Past, Present, and Future. International Materials Reviews, 2018, 63: 407-459.

[100] Matviychuk M M V, Savvakin D G. Synthesis of Highly Alloyed Titanium Alloys. Novi Mater. Tekhnol. Metallurg, 2010, 1: 81-84.

[101] 莫利亚尔·亚历山大,田金华,张莎莎,等.氢化钛粉制备钛及Ti-6Al-4V钛

合金粉末冶金工艺与性能研究. 南京航空航天大学学报, 2018, 50: 100-104.

[102] Benjamin J S. Dispersion Strengthened Superalloys by Mechanical Alloying. Metallurgical Transactions, 1970, 1: 2943-2951.

[103] Morales R, Aune R E, Seetharaman S, et al. The Powder Metallurgy Processing of Refractory Metals and Alloys. JOM, 2003, 55: 20-23.

[104] Johnson J L, Heaney D F, Myers N S. Metal Injection Molding (MIM) of Heavy Alloys, Refractory Metals, and Hardmetals//Heaney D F. Handbook of Metal Injection Molding. Cambridge: Woodhead Publishing, 2012:526-567.

[105] Shatta V. Powder Metallurgy: Sintered and Composite Materials [In Russian] Moscow: Metallurgy, 1983.

[106] Buckman R. New Applications for Tantalum and Tantalum Alloys. JOM, 2000, 52: 40-41.

[107] Zinkle S. Thermophysical and Mechanical Properties for Ta-8% W-2% Hf, Fusion Materials Semiann. Prog. Report for Period Ending, 1998, 27: 175-182.

[108] Sheftel E, Bannykh O. Niobium-Base Alloys. International Journal of Refractory Metals and Hard Materials, 1993, 12: 303-314.

[109] Krassenstein B. Niobium Based Alloy, C-103, Commonly Used for Aerospace Applications, Is Successfully 3D Printed by MTI. [2014-8-8]. https://3dprint.com/11409/niobium-c-103-alloy-3d-print.

[110] Klopp W D. Recent Developments in Chromium and Chromium Alloys. JOM, 1969, 21: 23-32.

[111] Encyclopedia of Inorganic Materials [In Russian]. GRUSE, 1977,9:819.

[112] Sims C T, Stoloff N, Hagel W. Superalloys II: Refractory Materials for Aerospace and Industrial Power Plants [Russian Translation]. Moscow: Metallurgiya, 1995.

[113] Karabasova Y S. New Materials [In Russian]. Moscow: Misis, 2002.

[114] Kim Y, Lee S, Kim E P, et al. Effects of ThO_2 on the Solid-State Sintering Behavior of W-Ni-Fe Alloy. International Journal of Refractory Metals and Hard Materials, 2011, 29: 112-116.

[115] Seltzer M, Wilcox B, Stringer J. The Oxidation Behavior of Ni-Cr-Al-$2ThO_2$ Alloys at 1093 ℃ and 1204 ℃. Metallurgical Transactions, 1972, 3: 2391-2401.

[116] Baranovskaya B N. Dispersion Hardening and Mechanical Alloying—New Ways of Creating High-Temperature Aviation Materials. Moscow: VIAM, 1994.

[117] Glazunov S G. Titanium Aluminides—The Basis of New Lightweight Superalloys for Aerospace Engineering. Moscow: VIAM, 1994.

[118] Skorokhod V. Powder Materials Based on Refractory Metals and Compounds. Kiev: Tekhnika, 1982.

[119] Vesenjak M, Borovinšek M, Fiedler T, et al. Structural Characterisation of Advanced Pore Morphology (APM) Foam Elements. Materials Letters, 2013, 110: 201-203.

[120] Arbuzova M V Z, Shmakov Y V, Andreyev D A. Aluminum Foam—A New Perspective Material for Aircraft, Cars and Ships. Tsvetnye Metally, 1997, 2: 62-65.

[121] Shalin R Y. Materials of Aerospace Engineering. Moscow: VIAM, 1994.

[122] Gribkov A N, Svetlov I L. Metal Composite Materials. Moscow: VIAM, 1994.

[123] Sapuan S M. Composite Materials. Boston: Butterworth-Heinemann, 2017.

[124] Awasthi S, Wood J L. Carbon/Carbon Composite Materials for Aircraft Brakes. Proceedings of the 12th Annual Conference on Composites and Advanced Ceramic Materials: Ceramic Engineering and Science Proceedings. Wiley Online Library, 1988: 553-559.

[125] Virgil'ev Y S, Kalyagina I. Carbon-Carbon Composite Materials. Inorganic Materials, 2004, 40: S33-S49.

[126] Reznik S, Prosuntsov P, Mikhailovskii K. Development of Elements of Reusable Heat Shields from a Carbon-Ceramic Composite Material 1: Theoretical Forecast. Journal of Engineering Physics and Thermophysics, 2019, 92: 89-94.

[127] Tinklepoe J R, Crandall W B. Cermets. Moscow: Publishing House of Foreign Literature, 1962.

[128] Schatt W, Wieters K P. Powder Metallurgy: Processing and Materials. European Powder Metallurgy Association, 1997.

[129] Fedorchenko I. Dictionary Reference on Friction, Wear, and Lubrication of the Parts of Machines. Kiev: Naukova Dumka, 1990.

[130] Fedorchenko I, Pugina L. Composite Sintered Antifriction Materials [In Russian]. Kiev: Naukova Dumka, 1980.

[131] Fedorchenko I, Frantsevich I, Radomyselskii I. Powder Metallurgy: Materials, Technology, Properties, Application. Handbook [In Russian] Naukova Dumka, 1985.

[132] Izrail Radomyselskii. Powder Metallurgy and Metal Ceramics, 2014, 53: 125-128.

[133] Bychkov S A, Yu O. Research of the Characteristics of Metal Fluoroplastic of Various Manufacturers for Elements of Aircraft Structures, Open Information and Computer Integrated Technologies, 2013, 59: 343-353.

[134] Rakovskii V, Silaev A, Khodkin V, et al. Powder Metallurgy of Heat-Resistant Alloys and Refractory Metals, 1974.

[135] Terry J E, Caras G J. Transpiration and Film Cooling of Liquid Rocket Nozzles. Redstone Scientific Information Center Redstone Arsenal Al, 1966.

[136] Mazyuk A L R, Ageyenko A V, Portyanko T N. Obtaining Capillary Structures of Contour Heat Pipes by the Method of Non-Shrinkable Sintering of Composite Powder Materials. Proceedings of the 29th International Conference on Industrial Composites. Kyiv, 2008:360-363.

[137] Dzneladze Z I, Shchegoleva R, Golubeva L. Powder Metallurgy of Steels and Alloys. Moscow: Metallurgiya, 1978.

[138] Vovk E, Deineka T, Doroshenko A, et al. Production of the Y3Al5O12 Transparent Nanostructured Ceramics. Journal of Superhard Materials, 2009, 31: 252-259.

[139] Garanin S, Rukavishnikov N, Dmitryuk A, et al. Laser Ceramic 1: Production Methods. Journal of Optical Technology, 2010, 77: 565-576.

[140] Chaynikova A S, Shchegoleva N E, Lebedeva Y E. Technological Aspects of Creating Radio-Transparent Glass-Crystalline Materials Based on High-Temperature Aluminosilicate Systems. Moscow: VIAM, 2015.

[141] Gnesin G, Dubok V, Braterskaya G. Sintered Materials for Electrical Engineering and Electronics: Handbook. Moscow: Metallurgiya, 1981.

[142] Cannesan N, Harris I. Aspects of Ndfeb HDDR Powders: Fundamentals and Processing. Bonded Magnets, 2003, 118: 13-36.

[143] Bulyk I I. Application of Hydrogen in the Production of Sintered Anisotropic Nanostructured Magnets from Alloys of Rare-Earth and Transition Metals. Materials Science, 2019, 54: 761-775.

[144] Gnesin G, Skorokhod V. Inorganic Materials Science: Encyclopedia in 2 Volumes [In Russian]. Kiev: Naukova Dumka, 2008.

[145] Guo J, Jiang H, Teng Y, et al. Recent Advances in Magnetic Carbon Nanotubes: Synthesis, Challenges and Highlighted Applications. Journal of Materials Chemistry B, 2021, 9: 9076-9099.

[146] Farhadi S, Farzaneh M, Kulinich S. Anti-Icing Performance of Superhydrophobic Surfaces. Applied Surface Science, 2011, 257: 6264-6269.

[147] Cao L, Jones A K, Sikka V K, et al. Anti-Icing Superhydrophobic Coatings. Langmuir, 2009, 25:12444-12448.

[148] Thomas S K, Cassoni R P, Macarthur C D. Aircraft Anti-Icing and De-Icing Techniques and Modeling. Journal of Aircraft, 1996, 33: 841-854.

[149] Qian H X, Xiong W H. Research on Nanocomposite Stealthy Materials. Aerospace Materials & Technology, 2002, 2: 2-11.

[150] Lyshevski S E. Nano- and Micro-Electromechanical Systems: Fundamentals of Nano- and Micro-engineering. CRC Press, 2018.

[151] Tiliakos N. MEMS for Harsh Environment Sensors in Aerospace Applications: Selected Case Studies//Kraft M, White N M. MEMS for Automotive and Aerospace Applications. Cambridge: Woodhead Publishing, 2013:245-282.

[152] Kraft M, White N M. MEMS for Automotive and Aerospace Applications. Elsevier, 2013.

[153] Najmon J C, Raeisi S, Tovar A. Review of Additive Manufacturing Technologies and Applications in the Aerospace Industry//Froes F, Boyer R. Additive Manufacturing for the Aerospace Industry. Elsevier, 2019: 7-31.

[154] Wohlers A. 3D Printing and Additive Manufacturing State of the Industry, Annual Worldwide Progress Report. Colorado: Associates Wohlers, 2016.

[155] Wohlers A. 3D Printing and Additive Manufacturing State of the Industry, Annual Worldwide Progress Report. Colorado: Associates Wohlers, 2017.

[156] PortolÉS L, JordÁ O, JordÁ L, et al. A Qualification Procedure to Manufacture and Repair Aerospace Parts with Electron Beam Melting. Journal of Manufacturing Systems, 2016, 41: 65-75.

[157] Mudge R P, Wald N R. Laser Engineered Net Shaping Advances Additive Manufacturing and Repair. Welding Journal, 2007, 86: 44-50.

[158] Walachowicz F, Bernsdorf I, Papenfuss U, et al. Comparative Energy, Resource and Recycling Lifecycle Analysis of the Industrial Repair Process of Gas Turbine Burners Using Conventional Machining and Additive Manufacturing. Journal of Industrial Ecology, 2017, 21: S203-S215.

[159] Dovbysh V M, Zabednov P V, Zlenko M A. Additive Technologies and Metal Products [In Russian]. Russia "NAMI" Caster Library, 2014, 9: 14-71.

[160] Frazier W E. Metal Additive Manufacturing: A Review. Journal of Materials Engineering and Performance, 2014, 23: 1917-1928.

[161] Chen G, Zhao S Y, Tan P, et al. A Comparative Study of Ti-6Al-4V Powders for Additive Manufacturing by Gas Atomization, Plasma Rotating Electrode Process and Plasma Atomization. Powder Technology, 2018, 333: 38-46.

[162] Yefimov N. Handbook of Non-Ferrous Metal Powders: Technologies and Applications. Elsevier, 2009.

[163] Franz H, Plochl L, Schimansky F P. Recent Advances of Titanium Alloy Powder Production by Ceramic-Free Inert Gas Atomization, Titanium 2008. Las Vegas: International Titanium Association, 2008.

[164] Entezarian M, Allaire F, Tsantrizos P, et al. Plasma Atomization: A New Process for the Production of Fine, Spherical Powders. JOM, 1996, 48: 53-55.

[165] Wosch E, Prikhodovski A, Feldhaus S, et al. Investigations on the Rapid Solidification of Steel Droplets in the Plasma—Rotating—Electrode—Process. Steel Research, 1997, 68: 239-246.

[166] Powell A. Electron Beam Centrifugal Atomization of Metal. [2019-5-15]. https://nanohub.org/resources/30540.

[167] Yap C Y, Chua C K, Dong Z L, et al. Review of Selective Laser Melting: Materials and Applications. Applied Physics Reviews, 2015, 2: 041101.

[168] Gibson I, Shi D. Material Properties and Fabrication Parameters in Selective Laser Sintering Process. Rapid Prototyping Journal, 1997, 3: 129-136.

[169] Kruth J P, Wang X, Laoui T, et al. Lasers and Materials in Selective Laser Sintering. Assembly Automation, 2003.

[170] Kruth J P, Mercelis P, Vaerenbergh J V, et al. Binding Mechanisms in Selective Laser Sintering and Selective Laser Melting. Rapid Prototyping Journal, 2005, 11: 26-36.

[171] Tillmann W, Schaak C, Nellesen J, et al. Hot Isostatic Pressing of IN718 Components Manufactured by Selective Laser Melting. Additive Manufacturing, 2017, 13: 93-102.

[172] Baumers M, Tuck C, Hague R. Selective Heat Sintering Versus Laser Sintering: Comparison of Deposition Rate, Process Energy Consumption and Cost Performance. SFF Proceedings, 2015, 1: 109-121.

[173] Murr L E, Gaytan S M, Ramirez D A, et al. Metal Fabrication by Additive Manufacturing Using Laser and Electron Beam Melting Technologies. Journal of Materials Science & Technology, 2012, 28: 1-14.

[174] Zhong Y, Rännar L E, Liu L, et al. Additive Manufacturing of 316L Stainless Steel by Electron Beam Melting for Nuclear Fusion Applications. Journal of Nuclear Materials, 2017, 486: 234-245.

[175] Murr L. Metallurgy of Additive Manufacturing: Examples from Electron Beam Melting. Additive Manufacturing, 2015, 5: 40-53.

[176] Graf B, Ammer S, Gumenyuk A, et al. Design of Experiments for Laser Metal Deposition in Maintenance, Repair and Overhaul Applications.

Procedia CIRP, 2013, 11: 245-248.

[177] Hauser C. Case Study: Laser Powder Metal Deposition Manufacturing of Complex Real Parts. TWI, 2014.

[178] Klimpel A, Lisiecki A, Janicki D. New Developments in the Process of the Laser Powder Surfacing. 23rd International Congress on Applications of Lasers & Electro-Optics, San Francisco, 2004.

[179] Grant P. Spray Forming. Progress in Materials Science, 1995, 39: 497-545.

[180] Thandalam S K, Ramanathan S, Sundarrajan S. Synthesis, Microstructural and Mechanical Properties of Ex Situ Zircon Particles ($Zrsio_4$) Reinforced Metal Matrix Composites (Mmcs): A Review. Journal of Materials Research and Technology, 2015, 4: 333-347.

[181] Grant P. Solidification in Spray Forming. Metallurgical and Materials Transactions A, 2007, 38: 1520-1529.

[182] Kellner T. An Epiphany of Disruption: GE Additive Chief Explains How 3D Printing Will Upend Manufacturing. GE Reports, 2017. https://www.ge.com/news/taxonomy/term/515?page=14.

[183] Bentur A. From Multiple to Singular—Consolidating Parts with Additive Manufacturing. LEO Lane, 2017.

[184] Heaney D F. Handbook of Metal Injection Molding. Cambridge: Woodhead Publishing, 2018.

[185] Nishiyabu K. Powder Space Holder Metal Injection Molding (PSH-MIM) of Micro-Porous Metals//Heaney D F. Handbook of Metal Injection Molding. Cambridge: Woodhead Publishing, 2012:349-390.